KU-677-534

Inorganic Complexes

David Nicholls
Senior Lecturer in Inorganic, Physical and Industrial Chemistry,
University of Liverpool

John Murray Albemarle Street London

© David Nicholls 1974

All rights reserved. No part of this publication may be reproduced, stored in a retrieval system, or transmitted, in any form or by any means, electronic, mechanical, photocopying, recording, or otherwise, without the prior permission of John Murray (Publishers) Ltd, 50 Albemarle Street, London W1X 4BD

Text set in 10/11 pt. Monotype Times New Roman, printed by photolithography, and bound in Great Britain at The Pitman Press, Bath

0 7195 2936 0

Contents

Preface

It is perhaps rather unfortunate that coordination compounds are called 'complex' compounds; to the new student this may imply that they are more difficult or 'complex' to understand than simple compounds. It may be for this reason that the subject is often inadequately dealt with in modern school textbooks. Students will probably find this text difficult at first; however, we cannot just ignore complex ions—they occur frequently in aqueous solution, and much understanding is to be gained once their presence is realized. This book is intended primarily for sixth-form students and teachers studying any of the newer syllabuses for the Advanced level of the General Certificate of Education. Some sections, notably Chapter 4, are beyond Advanced level but they are included for the benefit of teachers and others who wish for a brief introduction to the more theoretical aspects of the subject. It is hoped that parts of the text will also be useful in some first year university courses as well as in H.N.D. and similar courses.

The text is largely concerned with the interpretation of chemical reactions occurring in aqueous solution. In the author's opinion the understanding of complex compounds, and of their reactions, helps to systematize inorganic chemistry particularly at school level. For this reason, a rather academic approach is made to the subject. The importance of complexes in biological systems and as catalysts in organic chemistry is deliberately excluded as is the chemistry of the solid state. It is not that these topics are unimportant—on the contrary—but the rationalization of the common facts in inorganic chemistry has been considered more important in this short book. Some examples of the uses of particular complexes are given in the text.

The final chapter contains some suggested test-tube-scale experiments using common elements. Many of these reactions have been conducted in school laboratories over the years; it is the interpretations which now differ. Questions posed in the practical work are answered at the end of the chapter. This chapter can also therefore be used as a revision test although much of the joy and colour of transition metal chemistry is lost without practical work. A few ligands are introduced which are commonly used in coordination chemistry but which are not perhaps frequently found on the shelves in a school laboratory. In particular, EDTA, as the disodium salt, is relatively cheap and useful in illustrating the principles of coordination chemistry.

Liverpool, November 1971 *David Nicholls*

Introduction and Historical Development of Coordination Chemistry 1

What are complexes?

'Complexes' or as they are sometimes called 'coordination compounds' are as old as chemistry itself. They play a very important part in our lives today, and the study of them has contributed greatly to our understanding of the chemical bond and of inorganic chemistry as a whole. It is the purpose of this book to introduce the reader to complexes and to the understanding of chemistry which can be developed by their study.

First we must understand the meaning of the word complexes. We can say that a complex is formed when a number of ions or molecules combine with a central atom or ion to form an entity in which the number of ions or molecules directly attached to the central atom exceeds the normal covalency (oxidation state) of this atom. This formal definition serves only to show that it is difficult to say in a simple way what a complex is, yet the word presents no problem once experience of this type of compound is gained. Let us take some examples. Silver ions and chloride ions react in aqueous solution to give a precipitate of silver chloride:

$$Ag^+ + Cl^- \rightarrow AgCl$$

Silver chloride is not a complex because silver has a valency of one and is bonded to one chloride ion. If, however, we add aqueous ammonia to this precipitate, solution occurs because of the reaction:

$$AgCl + 2NH_3 \rightarrow [Ag(NH_3)_2]^+ + Cl^-$$

This new silver ion is a complex because the number of molecules bonded to the silver (two) is greater than the covalency (oxidation state) of the silver (one). Similarly, if we dissolve the silver chloride in concentrated hydrochloric acid, a complex is formed by means of the reaction:

$$AgCl + Cl^- \rightarrow [AgCl_2]^-$$

These two complexes are what are known as complex ions: $[Ag(NH_3)_2]^+$ carries a positive charge and is an example of a cationic complex, while $[AgCl_2]^-$ is negatively charged and is an example of an anionic complex. Such ions are exceedingly common—indeed in aqueous solution they are more common than simple ions. The reader will almost certainly have come across the blue solution of copper sulphate in water formed by the reaction:

$$\underset{\substack{\text{white}\\ \textit{(anhydrous)}}}{CuSO_4} + H_2O\,(\text{excess}) \rightarrow \underset{\text{blue}}{[Cu(H_2O)_6]^{2+}} + SO_4^{2-}(aq)$$

The blue colour is due to the complex cation which has six water molecules bonded to a Cu^{2+} ion. Indeed, most of the familiar blue copper(II) salts are complexes containing water molecules coordinated to the copper ions.

Although not so important in aqueous solution, there is a third type of complex compound, that is, the uncharged or neutral complexes. When boron trifluoride is reacted with ethoxyethane (ether) an 'addition' compound is formed:

$$BF_3 + Et_2O \rightarrow Et_2O \cdot BF_3$$

(Et denotes an ethyl group, C_2H_5.)

Many similar neutral complexes to this boron trifluoride etherate can be prepared, by direct reaction between two or more molecules, and these compounds will be discussed in later chapters.

Complexes, then, are formed by reaction of a metal ion or simple compound with other anions or neutral molecules. These anions or neutral molecules which bond to the central atom are called *ligands* or *donor molecules*. Some further examples of complex formation are listed in table 1.

'Simple' compound or ion	Ligand	Complex
Al^{3+}	F^-	$[AlF_6]^{3-}$
Al^{3+}	H_2O	$[Al(H_2O)_6]^{3+}$
Fe^{3+}	CN^-	$[Fe(CN)_6]^{3-}$
Co^{3+}	NO_2^-	$[Co(NO_2)_6]^{3-}$
Co^{2+}	Cl^-	$[CoCl_4]^{2-}$
BCl_3	Me_3N	$(Me_3N)BCl_3$
$SnCl_4$	Et_2O	$(Et_2O)_2SnCl_4$

(Me denotes a methyl group, CH_3.)

Table 1 Some complex compounds.

In the formation of a complex, the ligands can be considered to donate a pair of electrons to the metal or other central atom. Complexes are thus the products of coordinate bond formation. Because water is a donor molecule it is impossible to place a transition metal ion into aqueous solution without complex formation occurring; in the absence of any other competing ligand the complex formed is usually the hexaquo ion $[M(H_2O)_6]^{n+}$. Transition metal chemistry, and indeed the chemistry of many other metals in aqueous solution, is thus concerned with the reactions of this ion—and we shall consequently be examining these reactions in some detail.

An important distinction to be made is that between complexes and double salts. In double salts such as iron(II) ammonium sulphate, $FeSO_4{\cdot}(NH_4)_2SO_4{\cdot}6H_2O$, or potash alum, $KAl(SO_4)_2{\cdot}12H_2O$, the individuality of the constituent salts is maintained. Thus potash alum for example reacts chemically as a mixture of potassium sulphate and aluminium sulphate, and does not contain or show reactions of the $[Al(SO_4)_2]^-$ ion. In complex salts, the complex ion is present as a discrete entity and does not show the reactions of its components; potassium hexacyanoferrate(III), $K_3Fe(CN)_6$, for example, shows reactions of the $[Fe(CN)_6]^{3-}$ ion (as well as those of the K^+ ion) but not of the Fe^{3+} and CN^- ions. Very frequently double salts contain complex ions which are already present in one or both of the constituent salts. Thus potash alum contains the ion $[Al(H_2O)_6]^{3+}$ in the solid state and, like aluminium sulphate, gives this ion in aqueous solution. A more instructive way to write the formula for potash alum is therefore $K^+[Al(H_2O)_6]^{3+}(SO_4^{2-})_2{\cdot}6H_2O$.

Historical development

It is instructive to follow the historical development of coordination chemistry since much of the language and experimental techniques used can thereby be introduced. Of all elements, cobalt has played a very special role in our understanding of complexes. It was in 1798 that Tassaert observed that a solution of a cobalt(II) salt in aqueous ammonia becomes brown upon exposure to air, and that the colour changes to wine red when the solution is boiled. Later Frémy showed that the cobalt had been oxidized to cobalt(III) and that the new compound was associated with up to six molecules of ammonia as, for example, $CoCl_3{\cdot}6NH_3$. This compound alone was interesting because it was difficult to see why the two compounds, $CoCl_3$ and NH_3, both with their valencies satisfied, should combine to form a new stable species. Deeply coloured solutions were also obtained by dissolving copper(II) and nickel(II) salts in ammonia, and salts such as the purple $NiCl_2{\cdot}6NH_3$ and the deep blue $CuSO_4{\cdot}4NH_3{\cdot}H_2O$ were isolated. Subsequently, it was shown that a wide variety of compounds of cobalt(III) with ammonia (cobaltammines) could be isolated; it was the study of these and similar compounds of platinum that was to shed most light upon the structures of coordination compounds. At first these new compounds were named after their colours. Thus the yellow–orange $CoCl_3{\cdot}6NH_3$ was called 'luteocobaltic chloride', the purple $CoCl_3{\cdot}5NH_3$ 'purpureocobaltic chloride', and the green isomer of $CoCl_3{\cdot}4NH_3$ 'praseocobaltic chloride'. Other complexes were named after their discoverers: for example Magnus's green salt, $PtCl_2{\cdot}2NH_3$ (now formulated as

$[Pt(NH_3)_4][PtCl_4]$); and Zeise's salt, $PtCl_2 \cdot KCl \cdot C_2H_4$ (now formulated as $K[PtCl_3 \cdot C_2H_4]$). Today, complexes are named systematically and some of the internationally agreed rules for naming them are given at the end of this chapter.

It was in 1891 that Alfred Werner, at the age of 25, presented his first paper on the cobaltammines; twenty-two years later he received the Nobel prize for chemistry. Before summarizing Werner's postulates let us look at some of the experimental evidence upon which his theories were based.

Ionizable and non-ionizable chloride

When the three cobaltammines, $CoCl_3 \cdot 6NH_3$, $CoCl_3 \cdot 5NH_3$, and $CoCl_3 \cdot 4NH_3$, are treated with silver nitrate solution, they do not behave similarly. Whilst all the chloride ions in $CoCl_3 \cdot 6NH_3$ are precipitated immediately as silver chloride, the other two compounds precipitate only two and one chloride ions respectively. This information is summarized in table 2.

Complex	Number of Cl^- ions precipitated by $AgNO_3$	Present formulation
$CoCl_3 \cdot 6NH_3$	3	$[Co(NH_3)_6]^{3+} 3Cl^-$
$CoCl_3 \cdot 5NH_3$	2	$[Co(NH_3)_5Cl]^{2+} 2Cl^-$
$CoCl_3 \cdot 4NH_3$	1	$[Co(NH_3)_4Cl_2]^+ Cl^-$

Table 2 Ionizable chloride ions from cobaltammines.

The chloride ions precipitated by silver nitrate are known as the ionizable chlorides, while those not precipitated are the non-ionizable chlorides. Werner's conclusion, that these non-ionizable chlorides are covalently bonded to the cobalt, led to our present day formulations for these compounds as given in table 2. The number of ionizable chlorides is in each case that number of free chloride ions required to balance the charge on the cobalt cation as a whole.

Molar conductivity

The degree of ionization of a complex can also be determined by measuring the electrical conductivity of its aqueous solutions: the more ions a complex liberates into solution, the greater will be the conductivity. Substances which do not liberate ions when dissolved in water will give non-conducting solutions. In order for ready comparisons to be made it is convenient to measure molar conductivities (Λ). The molar conductivities of simple salts in dilute (10^{-3} M) aqueous solution are found to lie in the same region, that is about 120 Ω^{-1} cm^2 mol^{-1} for all salts, such as NaCl, producing two ions; salts producing three ions, such as $BaCl_2$, give molar conductivities in the region 260 Ω^{-1} cm^2 mol^{-1}, while salts giving four ions, such as $CeCl_3$, lead to molar conductivities of around 400 Ω^{-1} cm^2 mol^{-1}. Table 3 gives the molar conductivities of some platinum ammines with their present formulations in the final column.

Complex	Λ/Ω^{-1} cm^2 mol^{-1}	Number of ions	Present formulation
$PtCl_4 \cdot 6NH_3$	523	5	$[Pt(NH_3)_6]^{4+}4Cl^-$
$PtCl_4 \cdot 5NH_3$	404	4	$[Pt(NH_3)_5Cl]^{3+}3Cl^-$
$PtCl_4 \cdot 4NH_3$	229	3	$[Pt(NH_3)_4Cl_2]^{2+}2Cl^-$
$PtCl_4 \cdot 3NH_3$	97	2	$[Pt(NH_3)_3Cl_3]^+Cl^-$
$PtCl_4 \cdot 2NH_3$	0	0	$[Pt(NH_3)_2Cl_4]$

Table 3 Molar conductivities of some platinum ammines.

Werner's postulates

The interpretation of the experimental data of the kind we have just seen was made by Werner on the basis of several postulates. These can be summarized as follows.

1. Metals possess two types of valency: the primary (ionizable) valency and the secondary (non-ionizable) valency. Today we call these *oxidation state* and *coordination number* respectively.
2. The primary valencies are balanced by negative ions, secondary valencies are completed by negative ions and/or neutral molecules. Both valencies are satisfied.
3. The secondary valencies are directed to fixed positions in space about the central metal ion.

The first postulate introduces us to two terms which need clarification here. The primary valency we shall call oxidation state throughout. The oxidation state (or number) of an element in a compound can be defined formally as the number of positive charges that must be placed upon that element in order to neutralize the charges upon the other atoms. Let us take some examples. In copper chloride, $CuCl_2$, the copper must have a 2+ charge to neutralize the overall charge of 2− on the two chloride ions; hence its name copper(II) chloride. In potassium permanganate we have a charge of 1+ on the potassium and 4 × 2− on the oxygen atoms; thus manganese must be in the +7 oxidation state to make the salt electrically neutral. Note that oxidation state is a formalism: potassium permanganate does not actually contain Mn^{7+} ions nor is the manganese atom seven-valent. If there is no charge on the groups bonded to the metal then the metal is in oxidation state zero. In nickel carbonyl, $Ni(CO)_4$, for example, there is no charge on the carbon monoxide groups so in order to maintain neutrality there must be no charge on nickel. Again, it would not be realistic to regard the nickel as zero-valent since it is forming four bonds to the carbon monoxide groups. Thus the word 'valency' as applied to an element in a compound tends to be ambiguous and it is preferable to state its oxidation state and coordination number.

Werner's secondary valency is now universally called coordination number. The coordination number of an element in a compound is the number of ligands bonded directly to an atom of that element. In nickel carbonyl, the nickel has four CO groups bonded to each atom and hence the coordination number of the nickel is four. In $K_4Fe(CN)_6$ the iron atoms are six-coordinated by cyanide ions. Unfortunately, crystallographers use the term coordination number for ionic crystals; sodium ions in the sodium chloride lattice are surrounded octahedrally by six chloride ions and they are said to have a coordination number of six. As sodium chloride and other simple salts are not coordination compounds this ambiguity should not cause too much confusion.

We can now see how Werner's first two postulates apply to the cobaltammines (table 2). For each cobaltammine the oxidation state of the cobalt is $+3$ and the coordination number six and, in every case, both types of valency are satisfied. For the platinum ammines (table 3) the oxidation state of the platinum is $+4$ and the coordination number six in every case. We shall see that the coordination number of six is preferred for many metals, that is they prefer to have six ligands directly attached to the metal; these ligands are said to be in the coordination sphere of the metal. The ligands in the coordination sphere are bonded to the metal by coordinate bonds whereas the ions outside the coordination sphere are held ionically to the coordination sphere as a whole.

It was Werner's third postulate which, whilst being perhaps the most significant, was the most difficult to prove. If the bonds are fixed spatially then the existence of different isomers (isomers are compounds having the same empirical formula but different structures) can be predicted. The significance of this third postulate lies in its contribution to the development of inorganic stereochemistry. The fact that so many of the complexes known in Werner's time had a coordination number of six raised the important question of how the six ligands were arranged about the metal. At that time, direct methods of structure determination such as X-ray methods were not available and a choice between the various possible arrangements for six-coordination had to be made on the basis of the number of stereoisomers found and on the occurrence of optical isomers. It took Werner nearly twenty years to prove this postulate (and with it the proof of the octahedral structure for his six-coordinate compounds) via the resolution of an inorganic complex into its predicted optical isomers.

Nomenclature

It is very important that chemists the world over use the same name for each of the multitude of complex compounds. It was to this end that a Commission of the International Union of Pure and Applied Chemistry laid down, in 1958, a set of rules for naming these compounds unambiguously. These IUPAC rules are now in widespread use and some of the guide-lines will be given here.

The adoption of the Stock notation has greatly improved the naming of simple as well as complex compounds. In this notation the oxidation state of the metal is stated immediately after its name as a Roman numeral in parentheses; the old endings -ous, -ic, -yl, etc. thus disappear. The old problem to the beginning student of chemistry of whether 'ferrous' is di- or tri-valent iron now no longer arises—thus iron(II) chloride is clearly $FeCl_2$.

The cation is named first then the anion. The ligands are arranged before the central metal atom, negative ones first and then neutral ones; the number of each type of ligand is specified by Greek prefixes, mono-, di-, tri-, tetra-, penta-, hexa-, hepta-, and octa-. Thus:

$[Co(NH_3)_6]Cl_3$	hexamminecobalt(III) chloride
$[Co(NH_3)_5Cl]Cl_2$	chloropentamminecobalt(III) chloride
$[Co(NH_3)_4Cl_2]Cl$	dichlorotetramminecobalt(III) chloride

Note that it is not necessary to specify the number of chloride ions, that is to name $[Co(NH_3)_6]Cl_3$ as a trichloride, because, if the oxidation state is specified, the charge on the cation and hence the number of anions required to balance this charge are known. Note also that names of negatively charged ligands end in -o; thus we have bromo (Br^-), cyano (CN^-), nitro (NO_2^-). Neutral ligands keep their own names with the exception of ammonia which is called 'ammine', water which is called 'aquo', and carbon monoxide which is called 'carbonyl'.

If the complex is in an anionic form then the ending -ate is attached to the name of the metal and this is followed by the oxidation state of the metal. Thus

$(NH_4)_2TiCl_6$	ammonium hexachlorotitanate(IV)
$NaMn(CO)_5$	sodium pentacarbonylmanganate(−I)
K_2PtCl_4	potassium tetrachloroplatinate(II)

Some metals take their Latin name when in anionic form; these are

tin	stannate
lead	plumbate
copper	cuprate
silver	argentate
gold	aurate

For example

$K[Ag(CN)_2]$	potassium dicyanoargentate(I)

Neutral complexes are named in one word; for example:

$[Pt(NH_3)_2Cl_2]$	dichlorodiammineplatinum(II)
$[Co(NH_3)_3(NO_2)_3]$	trinitrotriamminecobalt(III)

Use is made of the prefixes bis-, tris-, tetrakis-, pentakis-, hexakis-, etc. to denote the number of ligand molecules when the ligand has the word di- or tri-, etc. at the beginning of its name, and the ligand is placed in brackets:

$SnCl_4(Et_2NH)_2$ tetrachlorobis(diethylamine)tin(IV)
$Ni(CO)_2(PPh_3)_2$ dicarbonylbis(triphenylphosphine)nickel(0)

(Ph denotes a phenyl group C_6H_5.)

These prefixes can also be useful where ambiguity might otherwise arise:

$[Fe(en)_3][Fe(CO)_4]$ tris(ethane-1,2-diamine)iron(II) tetracarbonylferrate(-II)
$BeCl_2(MeNH_2)_2$ dichlorobis(methylamine)beryllium(II)

(en denotes an ethane-1,2-diamine molecule, $H_2NCH_2CH_2NH_2$.)

Table 4 shows the systematic names of some common complexes. The systematic name has the disadvantage of being longer but it does enable the formula to be written down—which is more than we can say for 'nitroprusside'. With non-transition elements of invariable, or almost invariable, oxidation state there is no need to write the oxidation state—hence plain 'aluminate' in table 4.

Formula	Systematic name	Older name
Na_3AlF_6	sodium hexafluoroaluminate	cryolite
$K_3Fe(CN)_6$	potassium hexacyanoferrate(III)	potassium ferricyanide
$K_4Fe(CN)_6$	potassium hexacyanoferrate(II)	potassium ferrocyanide
$Na_2[Fe(CN)_5NO]$	sodium pentacyanonitrosyl-ferrate(II)	sodium nitroprusside
$Na_3Co(NO_2)_6$	sodium hexanitrocobaltate(III)	sodium cobaltinitrite

Table 4 Systematic names of some complexes.

The Formation of Complex Compounds: Donor–Acceptor Reactions 2

Lewis acids and bases

We have already seen that complexes can be regarded as being made up from a donor molecule or ligand and an acceptor molecule or ion. The development of this concept was due largely to an outstanding American professor—G. N. Lewis. In 1938 Lewis proposed definitions for the terms 'acid' and 'base' which were based upon electronic structure and hence he made these terms applicable, not only to aqueous solutions, but also to reactions in inert solvents and the gas phase. Lewis's definitions are:

Lewis acid—an electron-pair acceptor
Lewis base—an electron-pair donor

Lewis acids and bases thus react together through the formation of a coordinate bond. The result of this reaction is usually the formation of a complex, but simple neutralization reactions such as

$$H^+ + :\ddot{\underset{\cdot\cdot}{O}}:H^- \rightarrow H:\ddot{\underset{\cdot\cdot}{O}}:H$$

are still included. In order to show the usefulness of this classification the types of Lewis acids and Lewis bases are listed in tables 5 and 6.

Type	Examples
1 Positive ions	H^+, Ag^+, Cu^{2+}, Al^{3+}, Sn^{4+}
2 Molecules with unfilled octets	$BeCl_2$, BCl_3
3 Compounds, especially halides, in which the central atom may exceed its octet (i.e. has a vacant orbital of low energy)	$SiCl_4$, $GeCl_4$, $TiCl_4$, PCl_5

Table 5 Lewis acids (acceptors).

Type	Examples
1 Negative ions	OH^-, Cl^-, CN^-, SO_4^{2-}
2 Molecules with one or two lone pairs of electrons	NH_3, NMe_3, PPh_3, H_2O, Me_2S
3 Molecules containing carbon-to-carbon multiple bonds	$CH_2{=}CH_2$, $HC{\equiv}CH$, C_6H_6

Table 6 Lewis bases (ligands).

Note that the term Lewis acid is synonymous with 'acceptor' and 'electrophile', while Lewis base is synonymous with 'donor', 'nucleophile', or 'ligand'. Thus complex-forming reactions may equally well be called 'Lewis acid–base reactions', 'donor–acceptor reactions', or, for example, 'metal ion–ligand reactions'.

Now, since acids and bases react together we are able to make predictions of reaction behaviour between compounds. Obviously, positive ions are neutralized by negative ions with salt (but not necessarily complex) formation. Often the salt will, however, react with further base so that we can exemplify complex formation between type 1 Lewis acids and bases by:

$$Ag^+ + 2CN^- \rightarrow [Ag(CN)_2]^-$$
$$Co^{2+} + 4Cl^- \rightarrow [CoCl_4]^{2-}$$
$$Al^{3+} + 6F^- \rightarrow [AlF_6]^{3-}$$

The reactions of positive ions with molecules containing lone-pair electrons can be illustrated by:

$$Ag^+ + 2NH_3 \rightarrow [Ag(NH_3)_2]^+$$
$$Ni^{2+} + 6NH_3 \rightarrow [Ni(NH_3)_6]^{2+}$$
$$Al^{3+} + 6H_2O \rightarrow [Al(H_2O)_6]^{3+}$$

The strength of cations as acids or acceptors varies over a wide range. In general the acidity or coordinating ability increases with increase in charge on the cation. Thus Na^+ ions are weaker acids than Al^{3+} ions, and Sn^{2+} ions are weaker acids than Sn^{4+} ions. Whilst all cations attract strong bases such as water molecules to some extent, it is mainly the later transition metal ions which form stable complexes with unsaturated organic molecules; for example:

$$PtCl_4^{2-} + C_2H_4 \rightarrow [PtCl_3(C_2H_4)]^- + Cl^-$$

The reasons for the stability of this type of complex with particular metal ions need not concern us here: suffice it to say that these complexes of unsaturated organic molecules are very numerous and important.

The second type of Lewis acid concerns molecules with an unfilled octet. These molecules have a tendency to accept pairs of electrons from bases so as to complete their octet. In general the acidity of these molecules increases with increasing electronegativity (electron-withdrawing power) of the atoms or groups attached to the central atom. The more electronegative the attached groups the more a fractional positive charge is built up on the central atom and the more this atom is willing to receive an electron pair from a base. Thus the halides are strongly acidic while the hydroxides, such as $B(OH)_3$, are only weakly Lewis acidic. Let us consider the chlorides of beryllium, boron, and carbon:

```
             Cl  Cl        Cl
              ·· ··        ··
Cl:Be:Cl       B       Cl:C:Cl
               ··          ··
               Cl          Cl
```

Beryllium chloride is an acceptor molecule; as it is drawn here, it requires two pairs of electrons to complete its octet. The stoichiometry of complex formation will thus involve reaction between one molecule of $BeCl_2$ and two of a Lewis base. Typical tetrahedral complexes of beryllium are thus formed when the chloride reacts with ligands; for example:

$$BeCl_2 + 2Et_2O \rightarrow BeCl_2{\cdot}2Et_2O$$
$$BeCl_2 + 2Cl^- \rightarrow BeCl_4^{2-}$$

Boron trichloride is similarly acidic but will combine with only one molecule of a base to complete the octet; for example:

$$BCl_3 + Me_3N \rightarrow BCl_3{\cdot}Me_3N$$
$$BCl_3 + Cl^- \rightarrow BCl_4^-$$

Carbon tetrachloride, however, is not a Lewis acid; it already has a complete octet and so does not react under normal conditions with water, ether, chloride ions, etc. The reason for the stability of the octet with elements in the second row of the periodic table lies in the high energy of the 3d orbitals which are required for the formation of structures with five or six electron pairs around a central atom. The energy which must be put into a reaction before it will proceed is thus too great for the reaction to occur at room temperature and pressure.

With elements beyond the second row the energy of the d orbitals is not very far above that of the highest occupied orbital so that the octet rule no longer applies. Most commonly six rather than four electron pairs form a stable structure around these elements. Hence type 3 (table 5) of the Lewis acids comprises such halides as silicon tetrachloride. Reactions with bases normally yield six-coordinate complexes; for example:

$$SiF_4 + 2F^- \rightarrow SiF_6^{2-}$$

As a result of this acidity of the tetrahalides of silicon, germanium, and tin, they undergo reaction with water. The first step in the hydrolysis of these halides can be considered as the coordination of water molecules:

$$SiCl_4 + 2H_2O \rightarrow SiCl_4(OH_2)_2$$

(OH₂ above and below Si, coordinated by arrows; four Cl around Si)

In the presence of excess water, hydrolysis occurs by elimination of hydrogen chloride:

$$SiCl_4(H_2O)_2 + H_2O \rightleftharpoons SiCl_3(OH)(H_2O)_2 + HCl$$

until finally the equilibrium will be shifted right over to $Si(OH)_4{\cdot}2H_2O$ (hydrated silica) through the stages:

$$SiCl_3(OH)(H_2O)_2 + H_2O \rightleftharpoons SiCl_2(OH)_2(H_2O)_2 + HCl$$
$$SiCl_2(OH)_2(H_2O)_2 + H_2O \rightleftharpoons SiCl(OH)_3(H_2O)_2 + HCl$$
$$SiCl(OH)_3(H_2O)_2 + H_2O \rightleftharpoons Si(OH)_4(H_2O)_2 + HCl$$

The failure of carbon tetrachloride to hydrolyse is attributed to the high energy barrier to coordination of the water molecules in the pre-hydrolysis step.

For the sake of completeness we should mention that molecules possessing double bonds between unlike atoms often function as Lewis acids. In this category are CO_2, SO_2, and SO_3. After coordination of a water molecule a proton shift often occurs (although apparently not with SO_2):

$$\begin{array}{c} \mathrm{O} \\ \mathrm{O:\ddot{S}} \\ \ddot{\mathrm{O}} \end{array} + \begin{array}{c} \\ \mathrm{:\ddot{O}:H} \\ \mathrm{H} \end{array} \rightarrow \begin{array}{c} \mathrm{H} \\ \mathrm{O} \\ \mathrm{O:\ddot{\underset{\cdot\cdot}{S}}:OH} \\ \ddot{\mathrm{O}} \end{array}$$

(sulphur electrons only shown in SO_3).

The Lewis acid–base theory serves as a classification of donor and acceptor molecules: it enables us to comprehend a large section of chemistry and to make predictions of the likelihood of two chemicals reacting together. The words 'acid' and 'base' are not important. The organic chemist uses the Lewis ideas but calls the reactants electrophiles and nucleophiles, while the inorganic chemist tends to use the word ligand rather than Lewis base. Before discussing the types of ligand that exist we might just illustrate the kind of intelligent guess that can be made by using these ideas.

Let us consider two examples, the answers to which are, as far as the author knows, unknown. Firstly, will beryllium bromide react with dimethyl sulphide? Beryllium bromide is an acid; dimethyl sulphide, Me_2S, has a sulphur atom possessing two lone pairs of electrons and is therefore a base. We would, therefore, expect a reaction between these two species under suitable conditions; to give the tetrahedral octet the expected reaction is:

$$\mathrm{Br\!:\!Be\!:\!Br} + 2\mathrm{:\ddot{S}}\begin{array}{l} \diagup \mathrm{Me} \\ \diagdown \mathrm{Me} \end{array} \rightarrow \begin{array}{c} \mathrm{Br} \diagdown \quad \swarrow \mathrm{SMe_2} \\ \mathrm{Be} \\ \mathrm{Br} \diagup \quad \nwarrow \mathrm{SMe_2} \end{array}$$

Secondly, will tin(IV) chloride react with methyldiphenylphosphine, $MePPh_2$ ($Ph = C_6H_5$)? Again, tin(IV) chloride is a Lewis acid and the trivalent phosphorus atom contains a lone pair of electrons so that the reaction

$$\mathrm{SnCl_4} + 2\,\mathrm{:PMePh_2} \rightarrow \begin{array}{ccccc} & & \mathrm{Ph} & & \\ & \mathrm{Me} & | & \mathrm{Ph} & \\ & & \diagdown \, | \, \diagup & & \\ \mathrm{Cl} & & \mathrm{P} & & \mathrm{Cl} \\ & \diagdown & \downarrow & \diagup & \\ & & \mathrm{Sn} & & \\ & \diagup & \uparrow & \diagdown & \\ \mathrm{Cl} & & \mathrm{P} & & \mathrm{Cl} \\ & & \diagup \, | \, \diagdown & & \\ & \mathrm{Ph} & | & \mathrm{Me} & \\ & & \mathrm{Ph} & & \end{array}$$

would be expected to produce octahedrally coordinated tin. These then are just two predictions we might make; they are, of course, no substitute for doing the experiment and they may be wrong. Because the structure and nature of ligand molecules is so important in complex-forming reactions we shall now consider these in detail.

Ligands

How can we quickly decide whether or not a given molecule is capable of acting as a ligand? We need to recognize the atoms which possess lone-pair electrons in their more usual oxidation states. If we look at compounds of elements in the second row of the periodic table we can represent their structures as below, writing X to mean any other atom or group and two dots as a lone pair of electrons.

X—Be—X	BX_3	CX_4	$:NX_3$
acidic	acidic	neutral	basic

$X_2\ddot{O}:$	$X\ddot{F}:$	$:\ddot{Ne}:$
basic	neutral or acidic	neutral

We have already discussed the acidic properties of beryllium and boron compounds and the lack of donor or acceptor power shown by tetrahedral carbon. With nitrogen, in group V of the periodic classification, there are five valency electrons, only three of which are used in bonding to the atoms X, thus leaving two electrons—the lone pair. A trivalent nitrogen atom thus always possesses a lone pair of electrons and to illustrate the variety in X we can now list some examples of nitrogen ligands:

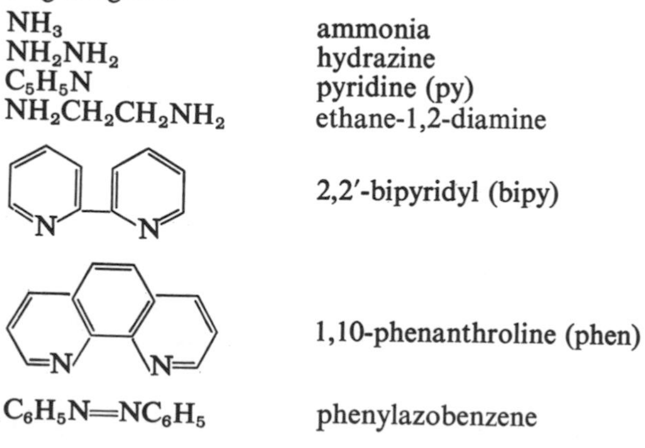

NH_3	ammonia
NH_2NH_2	hydrazine
C_5H_5N	pyridine (py)
$NH_2CH_2CH_2NH_2$	ethane-1,2-diamine
(structure)	2,2′-bipyridyl (bipy)
(structure)	1,10-phenanthroline (phen)
$C_6H_5N{=}NC_6H_5$	phenylazobenzene

Notice that the other group V elements, phosphorus and arsenic, also possess one lone pair when trivalent. Some phosphine and arsine ligands are:

PH_3	phosphine
PPh_3	triphenylphosphine
$C_6H_4(AsMe_2)_2$	*o*-phenylenebisdimethylarsine (diars)

In group VI we have oxygen, sulphur, and selenium each with two lone pairs of electrons when in oxidation state +2. Examples of ligands containing these atoms are:

C_2H_5OH	ethanol
$(C_2H_5)_2O$	ethoxyethane
$(CH_3)_2CO$	propanone
$(CH_3COCHCOCH_3)^-$	pentane-2,4-dionato
$(C_2H_5)_2S$	diethyl sulphide
C_4H_4S	thiophen

The conclusion we reach is that any organic molecule containing a neutral trivalent N, P, or As atom, or a neutral divalent O or S atom, can act as a ligand. This means that the number of complexes which can be synthesized is enormous—indeed, new compounds can readily be prepared by these 'acid–base' reactions in the school laboratory.

So far we have considered ligands containing only one donor atom, that is monodentate ligands. As can be seen from the list of nitrogen ligands there are many ligands which possess two donor sites and indeed others possessing three, four, five, or even six donor sites. These so-called multidentate ligands may act in one of two ways. Firstly, they may bond to a metal ion with all their donor atoms linked to the same metal ion; for example:

$$[CoCl_2(H_2NCH_2CH_2NH_2)_2]^+$$

The *trans*-dichlorobis(ethane-1,2-diamine)cobalt(III) cation

Here the ethane-1,2-diamine is acting as a *chelate* (crab-like) ligand.

Examples of anions which can chelate in this way are the pentane-2,4-dionato and ethanedioato anions, such as the trisethanedioatoaluminate ion:

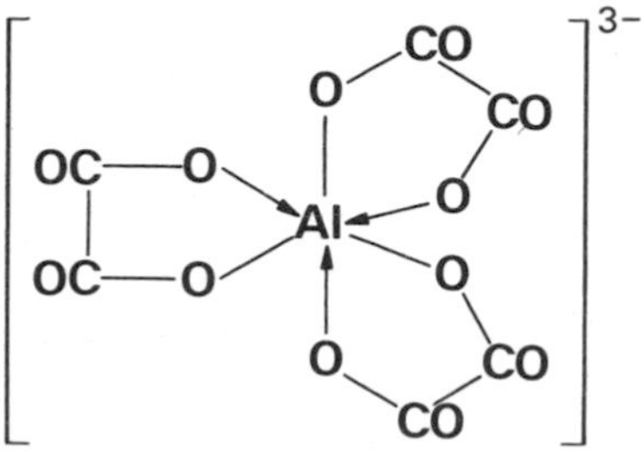

The most commonly used chelating ligand in aqueous solution is bis[di(carboxymethyl)amino]ethane (EDTA):

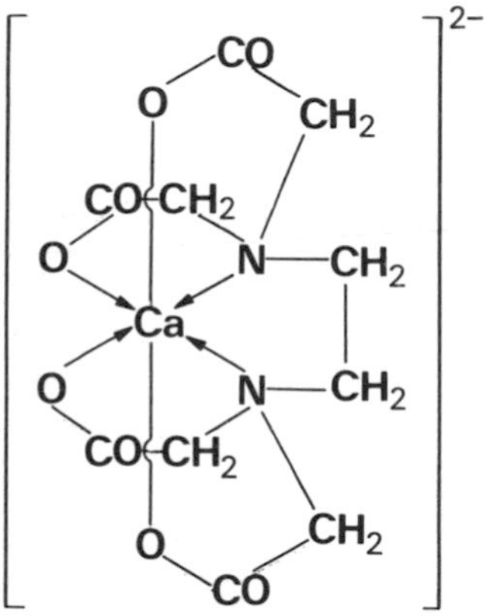

This possesses six donor atoms, four oxygens and two nitrogens as shown in its calcium complex above. The second way in which a multidentate ligand may be bonded in a complex is as a bridging group between two metal atoms; for example, in the polymeric cobalt(II) chloride hydrazine complex of repeating unit $CoCl_2(N_2H_4)_2$:

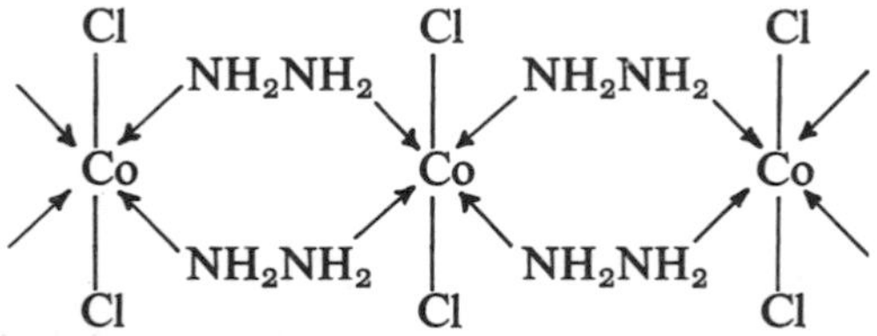

As a rule chelating ligands form more stable complexes than monodentate ligands with identical donor atoms.

The strength of Lewis bases depends upon many factors. Let us consider first the nature of the groups bonded to the donor atom. If we compare the strengths of the nitrogen bases, NMe_3, NH_3, and NF_3, the most usual order of basicity is

$$NMe_3 > NH_3 > NF_3$$

(It is difficult to be precise about Lewis base strength because it does vary with the nature of the Lewis acid.) With the very electronegative electron-withdrawing atoms (F, Cl, Br, I) the electron withdrawal along the N—X bond results in the lone pair of electrons on nitrogen being less accessible than in the case of NH_3. Indeed scarcely any complexes of the nitrogen trihalides acting as bases are known. In trimethylamine, the inductive effect of the methyl groups results in the lone pair of electrons on nitrogen being more accessible for bonding to an acid than in the case of ammonia. Secondly, let us compare the base strengths of a series of compounds with different donor atoms but with identical attached groups; for example:

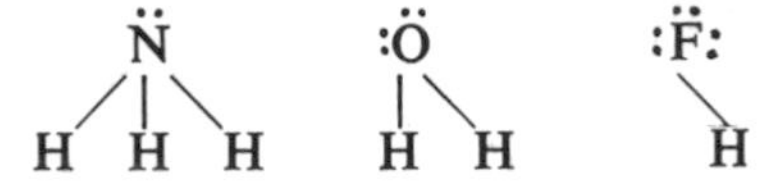

The base strength is observed experimentally to decrease from ammonia to hydrogen fluoride. In each case we have an octet around the hetero-atoms, but as we go from N to F the number of protons in the nucleus, and hence the nuclear charge, is increasing. Thus an increasingly strong pull on the unpaired electrons is felt as we go across this series so that these become less readily available for bonding to H^+ or any other Lewis acid. Another way of expressing this trend is to say that molecules with one lone pair are stronger bases than molecules having two lone pairs, and molecules with three lone pairs are scarcely basic at all.

Reactions in Aqueous Solution I: The Hexaquo Ion and its Acidity 3

The structure of water

The gaseous water molecule is bent with an H—O—H bond angle of about 104.5°. We can describe the bonding in the molecule in a variety of ways. Using electron-pair repulsion theory (see Chapter 5) we can say that the oxygen atom has six valency electrons, and each hydrogen atom one valency electron, so that there is an octet around oxygen. For four pairs of electrons, the tetrahedral structure is energetically preferred so that an H—O—H angle of about 109.5° might be expected. However, since the octet is composed of two bonding pairs and two lone pairs of electrons, the repulsions are not equal. The lone pairs exert greater repulsion upon each other and upon the bond pairs than do the bond pairs upon each other. Consequently, the bonding pairs of electrons are forced closer together and the bond angle is reduced from 109.5° to 104.5°. If we use hybrid orbitals to describe the water molecule, then we regard the oxygen atom as being sp^3 hybridized. There are six electrons for the four hybrid orbitals so that two of the orbitals are doubly filled and two singly filled. The singly filled orbitals overlap with the singly filled hydrogen 1s atomic orbitals, thus forming the bonds, while the doubly filled orbitals constitute the lone pairs of electrons. We can represent the water molecule diagrammatically thus:

doubly filled orbitals

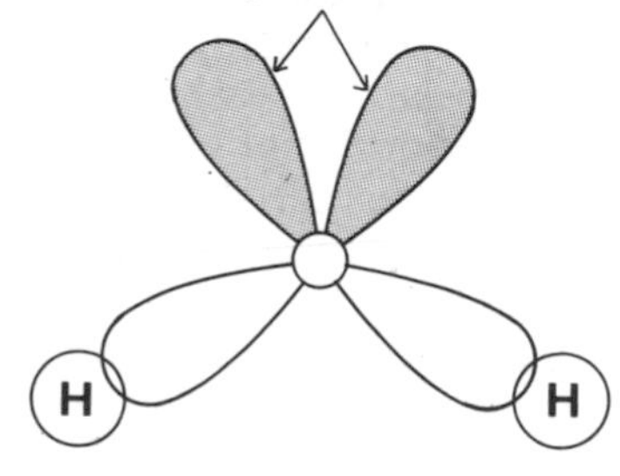

Liquid water is a polar solvent. In each water molecule, because of the electronegativity difference between hydrogen and oxygen, there is a charge separation. The hydrogen atoms have a fractional positive charge and the oxygen atoms a fractional negative charge. This charge separation is expressed and measured experimentally as the dipole moment which is usually represented as follows:

$O^{\delta-}$

$^{\delta+}H$ $H^{\delta+}$

↑ resultant dipole moment $= 6.16 \times 10^{-30}$ C m

This resultant dipole is the vector sum of all the moments in the various parts of the molecule. In water these moments arise not only from the polar nature of the O—H bonds (it is conventional to put the arrow head at the negative end of the dipole) but also from the presence of the two lone pairs of electrons. In liquid water, the water molecules do not exist in a free and uncombined state. The charge separation results in the molecules behaving as little electrical magnets—unlike poles attract, like poles repel. The resulting electrostatic bonding occurring throughout liquid water is known as 'hydrogen bonding'. The liquid can thus be represented as an array of water molecules joined together by hydrogen bonds (dotted lines):

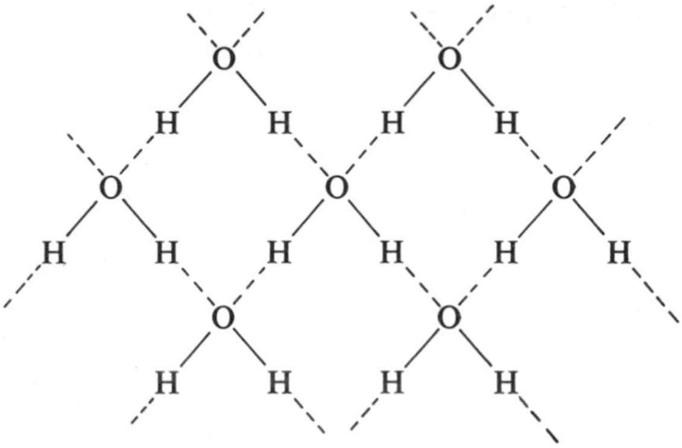

Notice that each oxygen atom is bonded to four hydrogens (not three as we might have expected on a simple dipole picture). Each lone pair of electrons on oxygen bonds to a different fractionally positively charged hydrogen atom. The coordination shape around each oxygen is thus roughly tetrahedral so that a three-dimensional (but not infinite) array is built up (not flat as it appears in the diagram above). Ice is known from X-ray studies to have these tetrahedral arrangements throughout in a very ordered structure. In liquid water the structure is less ordered but the local distribution around each oxygen remains tetrahedral. The extent of hydrogen bonding decreases as the temperature increases.

The nature of aqueous solutions

Now that we have considered the polar nature of water and seen how water molecules bond to each other, we need to consider what happens to ionic species when they are placed in aqueous solution. Let us consider a solution of a simple salt, for example sodium chloride. The water molecules can now form electrostatic bonds to the Na^+ and Cl^- ions and will do so in preference to bonding to other water molecules which have only partial charges on the hydrogen and oxygen atoms. Thus a 'coordination sphere' builds up around the sodium and chloride ions; we can represent this diagrammatically as in figure 1. The ions are said to be solvated.

Notice that the negative end of the water dipole is attracted to the cation and the positive end toward the anion. As well as this first, or so-called primary coordination sphere, further water molecules will be bonded more weakly in a secondary coordination sphere where they are further away from the cation. In a dilute solution at any one instant there will be many water molecules which are not bonded to any ion and which are therefore free to hydrogen-bond with other unattached molecules. The system is a very dynamic one, however, so that the

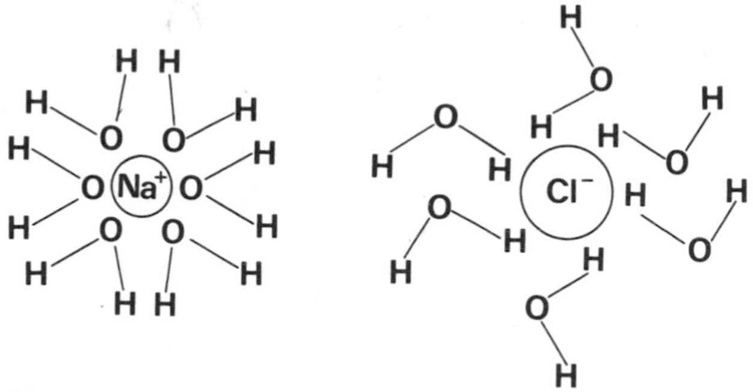

Figure 1 Solvation of a cation and an anion.

water molecules closest to the ions are exchanging continuously with other water molecules. This makes it very difficult to determine the exact number of water molecules which are directly bonded to the ions. With ions such as sodium and chloride, which do not have spectral or magnetic properties which change with coordination number, the number of first-sphere water molecules is not known with certainty. The problem is much more easily solved with transition metal ions and with highly charged ions which form coordinate (rather than solely electrostatic) bonds with water. We shall consider these cases shortly. At present let us classify the type of bonding which is present in the simple solution we have been considering. Because the interaction is occurring between an ion and the dipole of the water molecule the bond is called an ion–dipole bond, that is

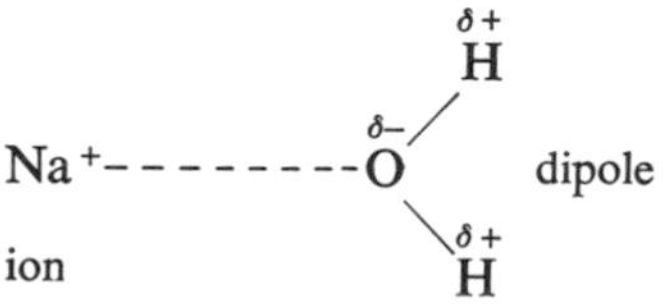

$$\underset{\text{ion}}{Na^{+}} \text{- - - - - - - -} \overset{\delta-}{O}\genfrac{}{}{0pt}{}{\diagup \overset{\delta+}{H}}{\diagdown \overset{\delta+}{H}} \quad \text{dipole}$$

Ion–dipole bonds are relatively weak: they are not as strong as ionic or covalent bonds. They may not always remain in the solid state; therefore sodium chloride crystallizes with the ionic lattice containing no water molecules. With larger and more highly charged cations, however, (as well as in many other sodium salts) salts often crystallize with water held by ion–dipole forces, that is water of crystallization. Such ion–dipole bonds are, however, readily broken and the water molecules are readily evolved upon heating the hydrates. Barium chloride, for example, crystallizes as the dihydrate $BaCl_2 \cdot 2H_2O$. In this salt the water molecules are bonded to the Ba^{2+} and Cl^- ions by ion–dipole forces. Upon heating to 100 °C the pure anhydrous salt is obtained:

$$BaCl_2 \cdot 2H_2O \rightarrow BaCl_2 + 2H_2O$$

The interaction between cations and water molecules is usually greater and therefore more important than that between anions and water molecules. The cation interacts with the lone-pair electrons on the oxygen atom, and the interaction is enhanced by the small size, and hence the high charge/radius ratio, of the cation. Anions are generally larger and hence the negative charge is more diffuse and the electrostatic interaction with water is weaker.

If an ionic solid is to dissolve in water, then the energy released by the hydration and dispersal of the ions must be greater than the energy of attraction between the ions in the solid lattice, that is the solvation energy must be greater than the lattice energy. Because two energy terms, enthalpy and entropy, contribute to the overall solvation energy, the process of solution may be exothermic or endothermic. Unfortunately, as both solvation and lattice energies are usually large quantities, we are frequently looking at a small difference between two very large quantities whose magnitudes cannot be determined with very great accuracy. We are not in a position, therefore, to predict solubilities, but when solution occurs it is certain that solvation has occurred. For the cations of groups I and II, and for simple anions, the enthalpy contribution to this solvation energy is derived from ion–dipole bonding. In view of the relative weakness of this type of bond we can, to some extent, ignore the presence of the bonded water molecules in chemical reactions since they do not interfere with the overall course of these reactions. With triply charged cations and with transition metal ions, however, the water molecules become held by much stronger forces and these coordinated water molecules cannot now be ignored if we are to understand the reactions of these ions in aqueous solution. It is the reactions of such aquo ions with which we shall be concerned for the rest of this chapter.

The hexaquo ion

In general, with tripositive ions of the main group elements, and with metal ions of the first transition series, water forms octahedral complexes of the general formula $[M(H_2O)_6]^{n+}$:

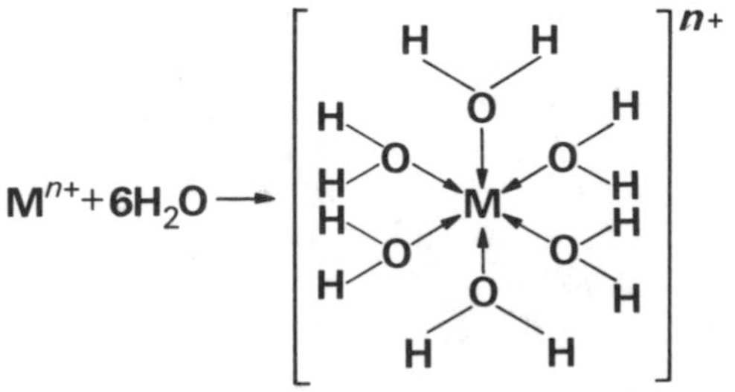

$$M^{n+}+6H_2O \longrightarrow [M(H_2O)_6]^{n+}$$

Four of these water molecules are in a plane containing the metal ion, one is above the plane, and one below it; thus all the O—M—O bond angles are 90°. In the equation, arrowheads are used on the O→M bonds to indicate that coordinate bond formation has occurred. Thus the pair of electrons in each bond can be regarded as having come originally from the oxygen atoms of the water molecules. We thus have six covalent bonds within the complex ion. This very important species $[M(H_2O)_6]^{n+}$ is called the hexa-aquo, hexaaqua, or hexaquo ion. Its presence is now well established both in the solid state and in solution.

X-ray crystallographic studies have shown that in the alums (empirical formula $M^IM^{III}(SO_4)_2 \cdot 12H_2O$), the trivalent metal ions (M^{III} = Al, Ga, Ti, V, Cr, Mn, Fe, Co) contain the octahedral $[M(H_2O)_6]^{3+}$ cations. In many transition metal compounds that we tend to regard as simple salts this complex hexaquo ion is again present. Thus $NiSO_4 \cdot 7H_2O$ is not nickel(II) sulphate heptahydrate (this name implies weak ion–dipole bonding) but is hexaquonickel(II) sulphate monohydrate or $[Ni(H_2O)_6]SO_4 \cdot H_2O$. Similarly, $FeSO_4 \cdot 7H_2O$ contains $[Fe(H_2O)_6]^{2+}$ and $CoSO_4 \cdot 7H_2O$ contains $[Co(H_2O)_6]^{2+}$ ions. Thus most of the transition metal salts on the laboratory shelves are in fact complexes and must be thought of as such if their chemistry is to be understood. Unfortunately, one cannot assume that any hydrated salt of a transition metal ion contains the hexaquo ion in the solid state. The six-coordinate octahedral structure is sometimes completed by a donor atom from an anion. Thus $CuSO_4 \cdot 5H_2O$ has a rather complicated structure (figure 2). Each copper atom is coordinately bonded to four water molecules in a plane and has two oxygen atoms from sulphate ions, one above and one below this plane, to complete a distorted octahedral array around copper (the Cu—OSO_3 bonds are longer than the Cu—OH_2 bonds). The fifth water molecule is hydrogen bonded to two sulphate oxygens

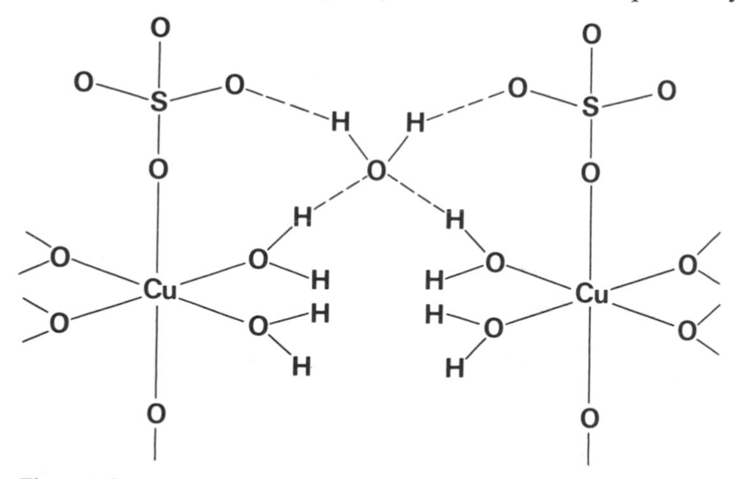

Figure 2 Structure of solid $CuSO_4 \cdot 5H_2O$ showing the position of the unique water molecule.

and two water hydrogens. Another deceptive example is $FeCl_2 \cdot 6H_2O$, which in the solid state contains *trans*-$[FeCl_2(H_2O)_4]$ units and not the hexaquo ion. Fortunately, in dilute aqueous solution even these compounds give hexaquo ions:

$$CuSO_4 + H_2O(\text{excess}) \rightarrow [Cu(H_2O)_6]^{2+} + SO_4^{2-}(aq)$$

and

$$FeCl_2 + H_2O(\text{excess}) \rightarrow [Fe(H_2O)_6]^{2+} + 2Cl^-(aq)$$

Evidence for the formation of hexaquo ions in solution is more easily obtained. The coloured ions of the transition elements have visible spectra which depend upon the stereochemistry of the metal ion. Thus cobalt(II) salts of non-complexing anions (such as NO_3^-, ClO_4^-) give pink aqueous solutions having spectra which are virtually identical to those of solid $CoSO_4 \cdot 7H_2O$ and $Co(ClO_4)_2 \cdot 6H_2O$, which are known from X-ray studies to contain $[Co(H_2O)_6]^{2+}$ ions. Even salts of the weakly complexing anions such as Cl^- and SO_4^{2-} give the same pink hexaquo ion in dilute solution. In high chloride ion concentrations some substitution of water molecules by chloride ions occurs so that the hexaquo ion is then no longer the major species present. We shall be discussing these substitution reactions in some detail in the next chapter.

The formation of hexaquo ions when metallic salts are dissolved in water is the general rule, then, and we must now consider the few exceptions to this rule. Obviously, elements in the second row (Li to Ne) which do not exceed the octet (d orbitals too high in energy) cannot form octahedral complexes by coordinate bonding. Beryllium salts thus give the tetrahedral $[Be(H_2O)_4]^{2+}$ ion in water. Higher co-ordination numbers than six are found amongst the second and third row transition elements and particularly in the lanthanides and actinides. Neodymium salts, for example, give the $[Nd(H_2O)_9]^{3+}$ ion; we shall not need to concern ourselves with such esoteric examples, however, as they are well beyond most school curricula at Advanced level. For reasons which are not yet fully understood silver(I) ions prefer linear two-coordination to six-coordination. Thus, it is more appropriate to consider aqueous solutions of silver salts as containing the $[Ag(H_2O)_2]^+$ ion rather than the $[Ag(H_2O)_6]^+$ ion. Copper(I) does not present this problem because of its disproportionation into $[Cu(H_2O)_6]^{2+}$ and Cu in water. Most of the discussion which follows, however, concerning general reactions of hexaquo ions is applicable also to these other ions which contain a different number of coordinated water molecules.

The acidity of aquo ions

The first general property of aquo ions that we shall examine is their Lowry–Brönsted acidity. All hexaquo ions are more or less acidic in the Lowry–Brönsted sense, that is they are capable of donating a proton to a base. Thus solutions of salts of di-, tri-, and tetravalent metal ions range from the almost neutral Fe^{2+} and Mn^{2+} to the very strongly acidic Fe^{3+} and Al^{3+}. The reaction involved in the production of hydrogen ions is as follows (we use a trivalent ion for simplicity):

$$\underset{\text{acid}}{[M(H_2O)_6]^{3+}} + \underset{\text{base}}{H_2O} \rightleftharpoons \underset{\text{base}}{[M(H_2O)_5(OH)]^{2+}} + \underset{\text{acid}}{H_3O^+}$$

The hexaquo ion protonates a water molecule thereby producing a hydroxo complex and the solvated hydrogen ion. As the reaction involves the donation of a proton by the aquo ion, then by definition the aquo ion is acting as an acid. An alternative description would be to say that the aquo ion has undergone hydrolysis. Of course, the hydrolytic reaction may continue in further stages, thus:

$$[M(H_2O)_5(OH)]^{2+} + H_2O \rightleftharpoons [M(H_2O)_4(OH)_2]^+ + H_3O^+$$
$$[M(H_2O)_4(OH)_2]^+ + H_2O \rightleftharpoons [M(H_2O)_3(OH)_3] + H_3O^+$$

The final product in this sequence is the hydrated metal hydroxide. Note that these reactions are all equilibria. If we add a base to such a solution the effect will be to consume H_3O^+ and thereby drive each of these equilibria toward the right-hand side. Ultimately, the metal hydroxide will precipitate. In the presence of added acid, however, the reverse will be true and the hexaquo ion will be the predominant species.

Just how acidic are solutions of metal ions? Table 7 lists the equilibrium constants (at room temperature) for the reactions

$$[M(H_2O)_6]^{n+} + H_2O \rightleftharpoons [M(H_2O)_5(OH)]^{(n-1)+} + H_3O^+$$

in the form of $pK_1 = -\log_{10}K_1$ where

$$K_1 = \frac{[H_3O^+][M(H_2O)_5(OH)^{(n-1)+}]}{[M(H_2O)_6^{n+}]}$$

It should be stated here that there is widespread disagreement in the literature between the values of K_1 obtained by different workers, since there are considerable problems in the experimental measurement of K_1. The data in the table are, however, useful in a semi-quantitative way. From the expression for K_1 it can be seen that a high value for K_1 indicates a high concentration of H_3O^+ and hence an acidic solution. Since K_1 varies from $10^{-0.1}$ to $10^{-10.9}$ for the ions chosen it can be seen that the lower values of pK_1 represent the more acidic solutions. We compare in table 7 the pK values for metal ions with the corresponding pK_a values for some more conventional acids. The data show that, whilst dipositive metal ions are not very acidic, the tripositive metal ions have an acidity which is of the same order of magnitude as that of ethanoic, methanoic, and phosphoric acids. Tetrapositive metal ions are extremely acidic in aqueous solution. For many of these, the first stage of the hydrolysis (of which K_1 is the equilibrium constant) lies well over

Ion	$pK_1(-\log_{10}K_1)$
Dipositive ions	
Mn^{2+}	10.9
Fe^{2+}	9.5
Zn^{2+}	9.0
Ni^{2+}	8.9
Cu^{2+}	8.0
Pb^{2+}	7.8
Co^{2+}	7.0
VO^{2+}	6.0
Tripositive ions	
Sc^{3+}	4.9
Al^{3+}	4.8
Cr^{3+}	3.9
V^{3+}	2.6
Ti^{3+}	2.6
Fe^{3+}	2.2
Tetrapositive ion	
Ce^{4+}	0.1
Other acids	pK_a
CH_3COOH	4.8
HCOOH	3.7
H_3PO_4	2.1

Table 7 Acidity of metal ions.

to the right-hand side so that the ions, $[M(H_2O)_6]^{4+}$, do not occur in significant amounts in aqueous solution. Thus, for example, vanadium(IV) compounds are normally found in aqueous solution as $[VO(H_2O)_5]^{2+}$; this species corresponds to $[V(OH)_2(H_2O)_4]^{2+}$, that is the second stage in the hydrolysis of $[V(H_2O)_6]^{4+}$ is complete. Nor is it usually possible to stabilize tetrapositive aquo ions by using very acidic solutions. The anions from mineral acids may undergo substitution reactions with aquo ions when they are present in high concentrations. Thus titanium(IV), for example, in concentrated hydrochloric acid forms $[TiCl_6]^{2-}$ rather than $[Ti(H_2O)_6]^{4+}$. There is the well-known experiment in tin chemistry in which metallic tin is treated with concentrated nitric acid with the resulting formation of a white solid. This solid used to be known as 'metastannic acid' but is now known as tin(IV) oxide, $SnO_2 \cdot xH_2O$ (x is probably 2). It seems curious to the student that a metal with a concentrated acid should produce an oxide. The oxide formation results from the extreme acidity of the $[Sn(H_2O)_6]^{4+}$ ion which is not known because it undergoes hydrolysis even in concentrated nitric acid (which of course contains water):

$$[Sn(H_2O)_6]^{4+} + 4H_2O \rightarrow [Sn(OH)_4(H_2O)_2] + 4H_3O^+$$

$$[Sn(OH)_4(H_2O)_2] \downarrow SnO_2 \cdot 2H_2O + 2H_2O$$

Other metals with a commonly occurring maximum oxidation state of +4 behave similarly in nitric acid solution.

The most important single factor then in determining the acidity of hexaquo ions is their charge. We can summarize what we have seen so far by saying that $[M(H_2O)_6]^{2+}$ is neutral or weakly acidic, $[M(H_2O)_6]^{3+}$ is strongly acidic, and $[M(H_2O)_6]^{4+}$ is extremely acidic. Let us consider why this is so by looking at the bonding situation in more detail.

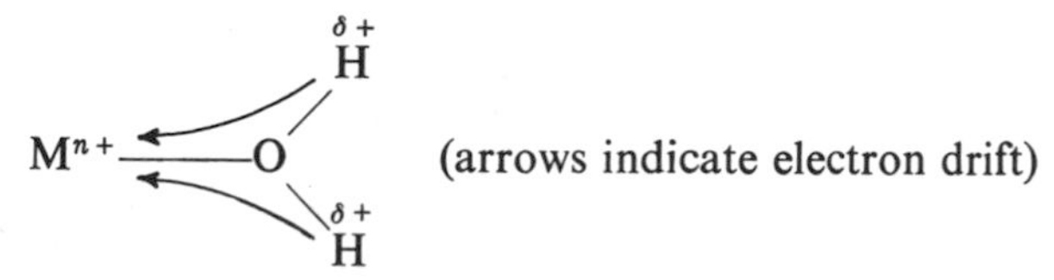

The higher the positive charge on the metal the greater the attraction it has for the electrons in the M—O bond. For highly charged ions, electrons drain away from the hydrogen atoms through the oxygen atom to the metal in an attempt to neutralize the positive charge. This has the effect of weakening the O—H bonds so that eventually the bond breaks because the proton can energetically bind more favourably to another water molecule. That is, the proton prefers to bind to the oxygen atom in HOH, a stronger Lewis base than $[M(H_2O)_5(OH)]^{(n-1)+}$. As the charge-to-size ratio of the metal ions becomes less (for example, going from Fe^{3+} to Fe^{2+}) the hexaquo ion becomes the more stable species and the acidity decreases correspondingly.

Before looking at the reactions of aquo ions as acids, we shall look at one hydrolytic reaction in more detail. The iron(III) cation behaves rather typically and in general what follows about its hydrolysis can be applied to other trivalent ions. It is well known that the addition of iron(III) chloride to water produces an acidic solution. The old textbooks used to say that the reaction occurring is:

$$FeCl_3 + 3H_2O \rightarrow Fe(OH)_3 + 3HCl$$

There is, however, no (or only a slight) precipitate of iron(III) hydroxide in such solutions and one is tempted to the conclusion that the hydrochloric acid produced according to the equation might have dissolved it. A much better and comprehensive description of this reaction involves hydrolysis of the hexaquo ion rather than of $FeCl_3$. We write the series of reactions as follows:

$$FeCl_3 + 6H_2O \rightarrow [Fe(H_2O)_6]^{3+} + 3Cl^-$$
$$[Fe(H_2O)_6]^{3+} + H_2O \rightleftharpoons [Fe(H_2O)_5(OH)]^{2+} + H_3O^+$$
$$[Fe(H_2O)_5(OH)]^{2+} + H_2O \rightleftharpoons [Fe(H_2O)_4(OH)_2]^+ + H_3O^+$$
$$[Fe(H_2O)_4(OH)_2]^+ + H_2O \rightleftharpoons [Fe(H_2O)_3(OH)_3] + H_3O^+$$

In iron(III) solutions, then, the position of the final equilibrium reached depends upon the pH of the solution. In the absence of added reagents the pH will be about 2 and while $[Fe(H_2O)_5(OH)]^{2+}$ and $[Fe(H_2O)_6]^{3+}$

are the major species there will be no precipitation (see table 7). If, however, we add a base, then by removal of H_3O^+ each equilibrium in turn is displaced towards the right until $[Fe(OH)_3(H_2O)_3]$ forms and is precipitated.

There are considerable complications to be added to this simple picture. Firstly, we chose iron(III) chloride because of its familiarity; it does, however, put chloride ions into the solution and these can and do to some extent form chlorocomplexes with iron(III) (see later). The $[Fe(H_2O)_6]^{3+}$ ion is violet (as in iron(III) alum) so that the yellow–brown solutions obtained in the hydrolysis reaction owe their colour to the hydroxo- and chloro-complexes. The addition of non-complexing acids such as perchloric acid to, for instance, $[Fe(OH)(H_2O)_5]^{2+}$ will restore the violet colour of the hexaquo ion. The processes involved in the precipitation of metal hydroxides by the addition of base are not quite so simple as we have so far implied. The hydroxo–aquo complexes may combine together in a process known as *olation*:

$$2[Fe(H_2O)_5(OH)]^{2+} \rightarrow [(H_2O)_4Fe\begin{array}{c} OH \\ \diagup\quad\diagdown \\ \diagdown\quad\diagup \\ OH \end{array}Fe(H_2O)_4]^{4+} + 2H_2O$$

This species with the hydroxo or *ol* bridges can react in the presence of added base as follows:

$$[(H_2O)_4Fe\begin{array}{c} OH \\ \diagup\quad\diagdown \\ \diagdown\quad\diagup \\ OH \end{array}Fe(H_2O)_4]^{4+} + 2H_2O \rightleftharpoons$$

$$2H_3O^+ + [(H_2O)_4Fe\begin{array}{c} O \\ \diagup\quad\diagdown \\ \diagdown\quad\diagup \\ O \end{array}Fe(H_2O)_4]^{2+}$$

The new ion species has *oxo* bridges and the process of its formation is known as *oxolation.* It is not known with certainty whether this species or the monoxo-bridged species is the more important:

$$[(H_2O)_4Fe\begin{array}{c} OH \\ \diagup\quad\diagdown \\ \diagdown\quad\diagup \\ OH \end{array}Fe(H_2O)_4]^{4+} \xrightarrow{H_2O} [(H_2O)_5Fe{-}O{-}Fe(H_2O)_5]^{4+}$$

Either oxo species can, however, now undergo consecutive hydrolysis, olation, and oxolation, and thus build up large aggregates which grow in size until the colloidal state is reached and precipitation occurs. The first stage of this process can be represented by the scheme:

$$[(H_2O)_5FeOFe(H_2O)_5]^{4+} \xrightarrow{H_2O} [(H_2O)_5FeOFe(H_2O)_4OH]^{3+} + H_3O^+$$

$$\downarrow \text{(i) olation, (ii) oxolation}$$

$$[(H_2O)_5FeOFe(H_2O)_4OFe(H_2O)_4OFe(H_2O)_5]^{6+}$$

The precipitate which is formed in the reaction of ammonia or sodium hydroxide with iron(III) solutions is amorphous to X-rays so that its exact nature is not known. It has a variable water content and is believed to consist of hydrated $FeO(OH)$. For simple equations, however, we shall find it convenient to represent these complex species as hydroxides, for example $Fe(OH)_3(aq)$ or $Fe(OH)_3(H_2O)_3$, bearing in mind that $FeO(OH)\cdot H_2O = Fe(OH)_3$ and $FeO(OH)\cdot 4H_2O = Fe(OH)_3(H_2O)_3$. This problem does not generally arise with divalent metal ions whose hydroxides exist as discrete compounds, as with $Fe(OH)_2$ or $Mn(OH)_2$.

Reactions of aquo ions with metals

We shall not be concerned here with the displacement of one metal by another according to the positions of the elements in the electrochemical series. These equations can easily be represented in terms of hexaquo ions; for example the addition of iron to copper(II) sulphate solution:

$$Fe + [Cu(H_2O)_6]^{2+} \rightarrow [Fe(H_2O)_6]^{2+} + Cu$$

In this example the hexaquocopper(II) ion is not sufficiently acidic to dissolve the iron metal. With solutions of tripositive ions, however, electropositive metals dissolve with effervescence. Consider the addition of magnesium turnings to a solution of aluminium sulphate. In this reaction effervescence occurs and eventually a white precipitate forms if excess of magnesium is present. The reactions occurring are:

$$[Al(H_2O)_6]^{3+} + H_2O \rightleftharpoons [Al(H_2O)_5(OH)]^{2+} + H_3O^+ \quad \text{and}$$

$$Mg + 2H_3O^+ \rightarrow Mg^{2+} + 2H_2O + H_2$$

(note that we write Mg^{2+} as unsolvated—it can also be written as $[Mg(H_2O)_6]^{2+}$—but as mentioned on page 20 this is unnecessary for cations in groups I and II). The gas evolved in the reaction is thus hydrogen. As the H_3O^+ ions react with the metal, the hydrolytic equilibrium is disturbed and the subsequent reactions occur with ultimate precipitation of the white 'hydroxide':

$$[Al(H_2O)_5(OH)]^{2+} + H_2O \rightleftharpoons [Al(H_2O)_4(OH)_2]^+ + H_3O^+$$

$$[Al(H_2O)_4(OH)_2]^+ + H_2O \rightleftharpoons [Al(H_2O)_3(OH)_3] + H_3O^+$$

Reaction of aquo ions with salts of weak acids

Under this heading we can discuss the reactions of hexaquo ions with alkali metal carbonates, sulphides, cyanides, ethanoates, and similar anions. In all these cases the more acidic hexaquo ions (those of M^{3+}), being stronger acids than those from which the anions are derived, displace the weaker acid. Thus the addition of sodium carbonate solution to iron(III) sulphate solution results in effervescence and the formation of a brown precipitate:

$$[Fe(H_2O)_6]^{3+} + H_2O \rightleftharpoons [Fe(H_2O)_5(OH)]^{2+} + H_3O^+$$
$$2H_3O^+ + CO_3^{2-} \rightarrow 3H_2O + CO_2$$

then

$$[Fe(H_2O)_5(OH)]^{2+} + 2H_2O \rightarrow Fe(H_2O)_3(OH)_3 + 2H_3O^+$$

With sodium carbonate, then, and a salt of a trivalent metal ion, carbonic acid is liberated, which decomposes into carbon dioxide and water. It is not possible therefore to prepare carbonates of iron(III), aluminium(III), vanadium(III), cerium(IV), etc. in aqueous solution. All attempts result in the precipitation of the metal hydroxides. Similarly, the addition of sodium sulphide solution to a solution of aluminium chloride yields a white precipitate of aluminium hydroxide with evolution of hydrogen sulphide:

$$[Al(H_2O)_6]^{3+} + H_2O \rightleftharpoons [Al(H_2O)_5(OH)]^{2+} + H_3O^+$$
$$2H_3O^+ + S^{2-} \rightarrow 2H_2O + H_2S$$

The reactions of trivalent metal ions with salts of weak acids are therefore very systematic and have predictable results.

With divalent metal ions, however, the results of these reactions are more complicated. Those ions of lower acidity than carbonic acid will yield metal carbonates when treated with, for example, sodium carbonate solution. Thus iron(II) carbonate is precipitated in pure form:

$$[Fe(H_2O)_6]^{2+} + CO_3^{2-} \rightarrow FeCO_3 + 6H_2O$$

With the more acidic M^{2+} ions, such as Ni^{2+}, Co^{2+}, and Cu^{2+}, the simple carbonates are not precipitated. The hydroxides are not precipitated either, but instead a 'half way' position in the hydrolytic equilibrium is reached and basic carbonates are precipitated. These basic salts are a characteristic feature of the chemistry of metals which form mildly acidic hexaquo ions. Their formulae are usually somewhat variable, that is salts of different stoichiometries are known; they are of the empirical form $xM(OH)_2 \cdot yMCO_3 \cdot zH_2O$ for divalent metals. We can understand their formation if we start from the hexaquo ion. The initial hydrolytic step occurs to some extent and is aided by the sodium carbonate solution:

$$[Cu(H_2O)_6]^{2+} + H_2O \rightleftharpoons [Cu(OH)(H_2O)_5]^+ + H_3O^+$$

Carbon dioxide is not usually evolved in sufficient quantity to be observed. The next stage in the hydrolysis would produce the hydroxide but as sodium carbonate is not sufficiently basic (or putting it another way, hexaquocopper(II) is not sufficiently acidic), this reaction effectively does not occur. We then have species such as $[Cu(OH)(H_2O)_5]^+$ in the solution which can grow into larger species by olation until these cations precipitate with whatever anion is present in an excess. Thus the addition of copper(II) sulphate solution to an excess of sodium carbonate solution precipitates the green basic carbonate; we can write its formation as follows:

$$2[Cu(OH)(H_2O)_5]^+ + CO_3^{2-} \rightarrow [Cu(OH)(H_2O)_5]_2CO_3$$
$$\textit{or}\ Cu(OH)_2 \cdot CuCO_3 \cdot 10H_2O$$

Obviously, with high sulphate concentrations some basic sulphate may precipitate: indeed, whole ranges of compositions are usually obtainable using a variety of different conditions. We shall not need to concern ourselves in such detail; we just need to note that pure carbonates are not expected from these slightly acidic aquo ions when they are treated with alkali metal carbonate solutions. Pure carbonates could, however, be precipitated if we could maintain the pH at such a point as to stabilize $[M(H_2O)_6]^{2+}$ and not allow hydrolysis to $[M(OH)(H_2O)_5]^+$. The addition of mineral acid would, of course, achieve this but this would also destroy added carbonate solution. The balance is achieved by using hydrogen carbonate solution. Either by the addition of sodium hydrogen carbonate, or by maintaining a pressure of carbon dioxide over the mixture of the metal salt and sodium carbonate, we can obtain the carbonates of these metals:

$$[Ni(H_2O)_6]^{2+} + 2HCO_3^- \rightarrow Ni(H_2O)_6CO_3 + H_2O + CO_2$$

Reactions of aquo ions with strong alkalis

The addition of ammonia or sodium hydroxide solutions to hexaquo ions usually results in the precipitation of the hydroxides. We can write the net equation, using, for example, a manganese(II) salt as:

$$[Mn(H_2O)_6]^{2+} + 2OH^- \rightarrow Mn(OH)_2(H_2O)_4 + 2H_2O$$

The four water molecules attached to $Mn(OH)_2$ as written above (there are three for M^{3+} ions) do not have great significance. Manganese(II) hydroxide exists as $Mn(OH)_2$ in the dry state as do many other hydroxides of metals in oxidation state +2. Whilst these hydroxides are suspended in the solutions, however, they almost certainly have water molecules associated with them, but the exact nature of these species is difficult to determine and not yet fully understood. Whether or not, in equations, we write such a species as hydrated, is thus to some extent a matter of personal preference. We include them here to show the relationship of the hydroxides in the hydrolytic equilibria and in the possible subsequent substitution reactions, as with, for example, hydroxide ions and ammonia.

The more acidic hexaquo ions give hydroxides which dissolve in an excess of sodium hydroxide solution—the hydroxides are said to be *amphoteric*. The well-known examples of this occur in the chemistries of aluminium and zinc. Various forms of aluminium 'hydroxide' are known, corresponding to formulae such as $AlO(OH)$ and $Al(OH)_3$. We shall for simplicity use the hydrated hydroxide in our example of amphoterism. In amphoterism we have the reversible equilibria:

$$[Al(OH)_2(H_2O)_4]^+ \underset{H_3O^+}{\overset{OH^-}{\rightleftharpoons}} [Al(OH)_3(H_2O)_3] \underset{H_3O^+}{\overset{OH^-}{\rightleftharpoons}} [Al(OH)_4(H_2O)_2]^-$$

The exact nature of the aluminate ion in solution is difficult to establish, but it is most reasonably written as $[Al(OH)_4]^-$ or $[Al(OH)_4(H_2O)_2]^-$ rather than AlO_2^- (note that $[Al(OH)_4]^-$ has the same empirical formula as $[AlO_2 \cdot 2H_2O]^-$). Similarly, zincates are written as $[Zn(OH)_4]^{2-}$. Many other metals exhibit amphoteric character: for example, cobalt(II) gives cobaltates(II), $[Co(OH)_4]^{2-}$; lead(II) gives plumbates(II); and titanium(IV) gives titanates(IV). Often rather extreme conditions (hot concentrated alkali for example) are required to dissolve the hydroxides in an excess of base. Copper(II) hydroxide will give a deep-blue solution in hot concentrated alkali, and iron(II) hydroxide is also soluble to some extent. The ferrate(II), $Na_2Fe(OH)_4$, can be obtained as fine blue–green crystals upon cooling the solution obtained by dissolving finely divided iron in boiling 50 per cent sodium hydroxide.

When the precipitation of hydroxides occurs in the presence of air, then oxidation by the air may occur. This is a common feature with the transition metal ions which have variable oxidation state. Manganese(II) hydroxide is a gelatinous white solid which rapidly darkens in the air as the manganese(III) hydroxide is formed; iron(II) hydroxide similarly turns from pale green to brown as oxidation to iron(III) occurs. With the transition elements it is a general rule that lower oxidation states are stabilized (with respect to oxidation) in acidic solution, while the higher oxidation states are readily obtained in alkaline solution. Since oxidation involves loss of electrons we can rationalize this general rule in a very elementary way. For a metal in oxidation state +2 we have the equations:

acidic solution $[M(H_2O)_6]^{2+} - e^- \rightarrow [M(H_2O)_6]^{3+}$
alkaline solution $[M(OH)_4(H_2O)_2]^{2-} - e^- \rightarrow [M(OH)_4(H_2O)_2]^-$

Obviously, it is easier to remove an electron from a species which is already negatively charged than to take an electron from a positive ion, and hence oxidation occurs more readily for the anionic species. These facts are supported by the values of the standard electrode potentials which for iron, for example, are as follows:

acidic solution $Fe^{3+} + e^- \rightarrow Fe^{2+}$; $E^\ominus = +0.77\ V$
alkaline solution $Fe(OH)_3 + e^- \rightarrow Fe(OH)_2 + OH^-$; $E^\ominus = -0.56\ V$

All the dipositive ions in the first transition series from Ti^{2+} to Co^{2+} undergo this aerial oxidation in alkaline solution; nickel(II) and copper(II) do not oxidize as they are already in their (normally) maximum oxidation state. Copper(I) in ammoniacal solution does, however, display this behaviour. This oxidation in ammoniacal solution also occurs with the dipositive ions and we shall give examples of this on page 35. It will be sufficient to mention here that for those aquo ions which do not readily undergo substitution by ammonia, the reaction stops at the hydroxide, $M(OH)_2$ or $M(OH)_3$, with no amphoterism. Thus, aluminium hydroxide is not soluble in aqueous ammonia because the concentration of hydroxide ions is insufficient.

Reactions in Aqueous Solution II: Substitution Reactions in the Hexaquo Ion 4

The stepwise replacement of water molecules

Perhaps the most general preparative route to complex compounds is to start from the most readily available complex—the hexaquo ion. By reaction with other ligands, the water molecules can be replaced in a stepwise fashion; this replacement may or may not go to completion in the presence of an excess of the ligand. Let us consider the reactions of ammonia with aquo ions. We can write the stepwise replacement of water molecules as follows:

$$
\begin{aligned}
&[M(H_2O)_6]^{n+} + NH_3 \rightleftharpoons [M(H_2O)_5(NH_3)]^{n+} + H_2O\\
&[M(H_2O)_5(NH_3)]^{n+} + NH_3 \rightleftharpoons [M(H_2O)_4(NH_3)_2]^{n+} + H_2O\\
&[M(H_2O)_4(NH_3)_2]^{n+} + NH_3 \rightleftharpoons [M(H_2O)_3(NH_3)_3]^{n+} + H_2O\\
&[M(H_2O)_3(NH_3)_3]^{n+} + NH_3 \rightleftharpoons [M(H_2O)_2(NH_3)_4]^{n+} + H_2O\\
&[M(H_2O)_2(NH_3)_4]^{n+} + NH_3 \rightleftharpoons [M(H_2O)(NH_3)_5]^{n+} + H_2O\\
&[M(H_2O)(NH_3)_5]^{n+} + NH_3 \rightleftharpoons [M(NH_3)_6]^{n+} + H_2O
\end{aligned}
$$

This replacement is readily observed because of the change in colour of the complex ions. Thus, if a solution of nickel(II) bromide is added to an excess of concentrated ammonia solution, the initially green nickel(II) solution becomes violet and the violet hexammine nickel(II) cation precipitates as the bromide:

$$[Ni(H_2O)_6]^{2+} + 2Br^- + 6NH_3 \rightarrow [Ni(NH_3)_6]Br_2 + 6H_2O$$

This hexammine ion is of course formed by all water-soluble nickel(II) salts but with most anions it stays in solution. A similar colour change is observed with copper(II) salts. When these are added to ammonia the colour change is from blue to deep blue–violet but in this case substitution does not go to completion in aqueous solution:

$$[Cu(H_2O)_6]^{2+} + 4NH_3 \rightleftharpoons [Cu(H_2O)_2(NH_3)_4]^{2+} + 4H_2O$$

Instead, the stepwise replacement proceeds until approximately four water molecules are replaced by ammonia. This is the equilibrium position in concentrated aqueous ammonia. If the ammonia solution is diluted, then the concentration of species such as $[Cu(H_2O)_3(NH_3)_3]^{2+}$ increases. The concentration of ammonia can only be increased by lowering the temperature or by using liquid ammonia—this of course yields the hexammine by direct reaction; for example:

$$CuBr_2 + 6NH_3 \rightarrow [Cu(NH_3)_6]Br_2$$

In aqueous solution, then, there is competition between the water and the ammonia molecules for the coordination sites on copper. If we entice precipitation from the solution obtained by treating copper(II) sulphate solution with an excess of ammonia (for instance, by the addition of ethanol) the resulting precipitate does not have quite the expected composition, that is $[Cu(NH_3)_4(H_2O)_2]SO_4$. Instead, we obtain $[Cu(NH_3)_4(H_2O)]SO_4$, which has a square pyramidal structure in which the four ammonia molecules are in a square plane around the copper, and the water molecule is at the apex of the pyramid. This illustrates a general point that complexes isolated from solution do not *necessarily* contain the species present in the solution although this is frequently the case.

Stability constants

These substitution reactions can be studied semi-quantitatively with the help of the stepwise equilibrium constants. In the ammonia-substitution reactions which we have been considering we can write

$$K_1 = \frac{[M(H_2O)_5(NH_3)^{n+}]}{[M(H_2O)_6^{n+}][NH_3]}$$

$$K_2 = \frac{[M(H_2O)_4(NH_3)_2^{n+}]}{[M(H_2O)_5(NH_3)^{n+}][NH_3]}$$

$$K_3 = \frac{[M(H_2O)_3(NH_3)_3^{n+}]}{[M(H_2O)_4(NH_3)_2^{n+}][NH_3]}$$

and so on. Note that the water molecules involved in these reactions are not included in the equilibrium constant; the activity of pure water is conventionally defined as unity. These stepwise constants are known as formation or *stability constants*. For an overall reaction we may be more interested in the overall stability constant which is usually denoted by the symbol β. In the formation of a metal hexammine, then, we write

$$\beta_6 = \frac{[M(NH_3)_6^{n+}]}{[M(H_2O)_6^{n+}]\,[NH_3]^6}$$

The β and K values are related in a simple way:

$$\beta_6 = K_1K_2K_3K_4K_5K_6$$

These stability constants have been measured experimentally for a large number of metal ions with a wide variety of ligands.* Note, however, that these constants relate to the thermodynamics of the substitution processes and tell us nothing about the rate at which equilibrium is attained. Some typical stability constants are given in table 8.

Metal ion	Ligand	log K_1	log K_2	log K_3	log K_4	log K_5	log K_6	log β
Cu^{2+}	NH_3	4.2	3.5	2.9	2.1			12.6
Ni^{2+}	NH_3	2.8	2.2	1.7	1.2	0.8	0.03	8.7
Cu^{2+}	en	10.6	9.1					19.6
Ni^{2+}	en	7.5	6.2	4.3				18.0
Cu^{2+}	EDTA	18.8						18.8
Ni^{2+}	EDTA	18.6						18.6

Table 8 Some stability constants of metal complexes.

In view of the magnitude of these constants they are usually quoted as logarithms; thus the equilibrium constant for the reaction

$$[Ni(H_2O)_6]^{2+} + 3en \rightarrow [Ni(en)_3]^{2+} + 6H_2O$$

is 1×10^{18}. The log β values in table 8 correspond to the products of the K values listed for each reaction. The blank spaces are those for which the stability constants are very low or negative; for example, for $[Cu(NH_3)_5(H_2O)]^{2+}$, log K_5 is about -0.5.

The general trends to be noted from table 8 are that the successive stability constants decrease steadily in any particular system, and that chelating ligands (such as ethane-1,2-diamine and EDTA) form more stable complexes than monodentate ligands having the same donor atom. The first effect is largely a statistical one—there is a greater probability of exchanging a water molecule in $[Ni(H_2O)_6]^{2+}$ than there is in $[Ni(H_2O)_5(NH_3)]^{2+}$, and so on. The second effect is often termed the *chelate effect*. The principal contributing factors to this extra stability of chelates are the inherent stability of five- and six-membered rings, and the significantly greater entropy change upon substitution by the chelate ligand as compared to that by a monodentate ligand. A comparison

*A compilation of such data is to be found in *Stability Constants*, special publication No. 17 of the Chemical Society (London 1964), and *Supplement No. 1* to this which is special publication No. 25 of the Chemical Society (London 1971).

of the reactions

$$[Ni(H_2O)_6]^{2+} + 6NH_3 \rightarrow [Ni(NH_3)_6]^{2+} + 6H_2O$$

and

$$[Ni(H_2O)_6]^{2+} + 3en \rightarrow [Ni(en)_3]^{2+} + 6H_2O$$

shows that for the ammonia reaction seven 'particles' produce another seven 'particles'. With the bidentate ligand, however, seven particles are produced from four. In this case, then, there is an increase in the randomness of the system and hence an increase in the entropy. Since entropy and the equilibrium constant are related through

$$\Delta G^{\ominus} = \Delta H^{\ominus} - T\Delta S^{\ominus} = -RT \ln K$$

the increase in entropy change for chelate ligand reactions is largely responsible for the increase in the equilibrium or stability constant K.

With the hexadentate ligand, EDTA, only one molecule of the ligand is required for each metal ion in order for the metal to remain six-coordinate (see Chapter 2) so that six water molecules are liberated. This ligand thus forms very stable complexes and finds many uses in analytical and industrial chemistry. In volumetric analysis a very large number of metal ions can be titrated directly with EDTA using weakly complexing dyes as indicators. As a water softener, EDTA added to water leaves no free calcium or magnesium ions to precipitate with soaps. Reactions of metal hexaquo ions are not therefore observable in the presence of EDTA since the aquo ions are no longer present.

Many ligands displace all the water molecules from aquo ions and we shall discuss numerous examples of such displacements shortly. Because the reactions are carried out in aqueous solution, however, there is competition between the water molecules and the ligand for the coordination sites on the metal ion. It is often necessary, therefore, to use a large excess of the ligand in order to effect complete substitution. The copper(II) hexaquo ion is rather exceptional. Either four monodentate ligands or two bidentate ligands readily replace four water molecules, but the last two are very difficult to remove in aqueous solution. The reasons for this difference, as compared to other divalent ions such as Co^{2+} and Ni^{2+}, are connected with the Jahn–Teller effect (Chapter 5). The $[Cu(H_2O)_6]^{2+}$ ion is not a perfectly octahedral ion; instead, it is distorted to a tetragonal structure with two Cu—OH_2 bonds longer than the other four (which are in the square plane). It is for this reason that many authors in elementary texts have used the formulae $[Cu(H_2O)_4]^{2+}$ and $[Cu(NH_3)_4]^{2+}$ for copper(II) complexes; these formulae, apart from being poor approximations to the truth, can be very misleading.

So far we have assumed direct substitution to occur with no side-reactions. Whilst this is realistic for most ligands, ammonia is somewhat exceptional in that it acts as a strong Lowry–Brönsted base as well as a Lewis base. We do, therefore, have the hydrolytic equilibria of the

$[Cu(H_2O)_6]^{2+}$ and similar ions disturbed so that the slow addition of ammonia to such ions in fact precipitates the hydroxides. For the strongly acidic ions, and those metals which prefer to bind to oxygen donors rather than nitrogen, no complexes with ammonia are formed. The hydroxides of the divalent metal ions, however, frequently dissolve when an excess of ammonia is added, and the respective point on the substitution equilibria is then reached (as though intermediate hydroxide precipitation had occurred). Another major reaction which commonly occurs along with substitution is oxidation. Again, as it is ammonia which produces an alkaline medium, and as oxidation occurs more readily under alkaline conditions, it is with ammonia in particular that oxidation by the air occurs. Finally we have the possibility of stereochemical change—for instance octahedral six-coordinate to tetrahedral four-coordinate—occurring during the substitution process. We shall now exemplify all these reaction types by considering the chemistry of many elements.

Substitution by neutral ligands

Neutral or uncharged ligands generally substitute into aquo ions to give complexes of the same coordination number as the starting aquo ion. Table 9 gives a summary of the reactions occurring between metal aquo ions (coordinated water molecules are omitted for clarity) and an excess of ammonia. It is impossible to cover all possibilities in a table of this form, and the table must be used in conjunction with the text. The distinction between metal ions which do form water-stable ammonia complexes and those which do not is somewhat arbitrary. From

Metal ion	Species formed in the absence of air	Species formed upon oxidation by air
Ions forming ammonia complexes		
Cr^{3+}	$[Cr(NH_3)_6]^{3+}$	no oxidation
Co^{2+}	$[Co(NH_3)_6]^{2+}$	$[Co(NH_3)_6]^{3+}$
Ni^{2+}	$[Ni(NH_3)_6]^{2+}$	no oxidation
Cu^{2+}	$[Cu(NH_3)_4(H_2O)_2]^{2+}$	no oxidation
Cu^{+}	$[Cu(NH_3)_2]^{+}$	$[Cu(NH_3)_4(H_2O)_2]^{2+}$
Ag^{+}	$[Ag(NH_3)_2]^{+}$	no oxidation
Zn^{2+}	$[Zn(NH_3)_4]^{2+}$	no oxidation
Ions not forming ammonia complexes		
Al^{3+}	$Al(OH)_3$	no oxidation
Ti^{3+}	$Ti(OH)_3$	$TiO_2 \cdot xH_2O$
V^{3+}	$V(OH)_3$	$VO(OH)_2$
VO^{2+}	$VO(OH)_2$	no oxidation
Fe^{2+}	$Fe(OH)_2$	$Fe(OH)_3$
Fe^{3+}	$Fe(OH)_3$	no oxidation
Mn^{2+}	$Mn(OH)_2$	$Mn(OH)_3$

Table 9 The reactions of an excess of aqueous ammonia with metal ions.

the preceding discussion the reader will be aware that many ammines (as ammonia complexes are called) are preceded by metal hydroxide formation; the rate and extent of dissolution of the hydroxide in an excess of ammonia varies widely from metal to metal. Thus chromium(III) hydroxide is rather slow to dissolve in an excess of ammonia, but deep-violet solutions are eventually formed. A vast chemistry of chromium(III) ammines is known: the simpler ones having compositions ranging from $[Cr(NH_3)_6]^{3+}$ to $[Cr(NH_3)_2(H_2O)_4]^{3+}$ are all known, as well as many more complicated species.

The addition of ammonia to a solution of a cobalt(II) salt starts off a sequence of reactions which ends with the formation of the celebrated cobaltammines which occupied so much of Werner's research. Initially, a blue precipitate of the cobalt(II) hydroxide forms (note that a pink form of $Co(OH)_2$ also occurs!) which dissolves in an excess of ammonia to form a pale straw-coloured solution. Left to stand in the air the solution turns darker brown and brown colours can be seen developing at the air–solution interface. The process can be speeded up by bubbling air through the solution or by adding a little hydrogen peroxide. Finally, a dark red–brown solution is obtained. The exact nature of the species present in this solution is still rather complicated. In order to isolate the crystalline cobaltammines, using chloride ions, for example, as the anion, it is necessary to add extra chloride ions in the form of ammonium chloride, the equation being

$$4[Co(H_2O)_6]Cl_2 + 4NH_4Cl + 20NH_3 + O_2 \rightarrow 4[Co(NH_3)_6]Cl_3 + 26H_2O$$

The orange–yellow hexamminecobalt(III) chloride is formed, particularly if a charcoal catalyst is used. In the absence of charcoal, and using hydrogen peroxide instead of air, the aquopentammine species predominates; this can be readily converted to the red chloropentammine cobalt(III) chloride by treatment with concentrated hydrochloric acid:

$$2[Co(H_2O)_6]Cl_2 + 2NH_4Cl + 8NH_3 + H_2O_2 \rightarrow 2[Co(NH_3)_5(H_2O)]Cl_3 + 12H_2O$$

$$[Co(NH_3)_5(H_2O)]Cl_3 \rightarrow [Co(NH_3)_5Cl]Cl_2 + H_2O$$

The oxidation of copper(I) is similar. An insoluble copper(I) salt must be used (or else disproportionation occurs, as with Cu_2SO_4) to prepare copper(I) ammines. Copper(I) chloride resembles silver chloride in dissolving in ammonia solution to give a colourless solution:

$$CuCl + 2NH_3 \rightarrow [H_3N - Cu - NH_3]^+ + Cl^-$$

$$AgCl + 2NH_3 \rightarrow [H_3N - Ag - NH_3]^+ + Cl^-$$

In both cases the linear complex ions are formed. As usually performed with copper(I) chloride slightly contaminated with copper(II), and ammonia solution containing dissolved oxygen, the CuCl–NH_3 reaction gives a pale-blue solution. As with cobalt(II), this solution can be seen to darken and this process can be speeded up by agitating the solution in air. Eventually, all the copper is oxidized to copper(II) and the deep violet–blue diaquotetrammine copper(II) ion is present:

$$2[Cu(NH_3)_2]^+ + 4NH_3 + 5H_2O + \tfrac{1}{2}O_2 \rightarrow 2[Cu(NH_3)_4(H_2O)_2]^{2+} + 2OH^-$$

Note that in all the ammines listed in table 9 the coordination number around the metal is the same as in the corresponding aquo ions (with the possible exception of the Cu^+–H_2O system which cannot be studied). Thus chromium(III), cobalt(II), nickel(II), and copper(II) remain six-coordinate, copper(I) and silver(I) have their preferred coordination number of two, and zinc prefers the tetrahedral four-coordination in $[Zn(H_2O)_4]^{2+}$ and $[Zn(NH_3)_4]^{2+}$ (although six-coordination is not uncommon with this element).

Those species which predominantly undergo base-hydrolysis when treated with aqueous ammonia are also listed in table 9 with their oxidation products. We have already discussed these reactions; the hydroxides are written in the table for simplicity, but their complex nature has also been indicated previously. Under rather special conditions some of these metal ions can be made to give ammines in aqueous solution. The hexammine iron(II) bromide, $[Fe(NH_3)_6]Br_2$, for example, can be prepared by saturating a solution of iron in hydrobromic acid (under an atmosphere of hydrogen) with ammonia; the pale blue–grey crystals decompose upon exposure to air. All the ions listed in the bottom half of table 9 form ammines when water is absent, that is, either in liquid ammonia or by reaction of ammonia gas with an anhydrous salt of the metal ion. These ammines react immediately in water or aqueous ammonia however to give the products stated in table 9. The reactions of these ammines with water can be regarded as proceeding via the hexaquo ions which then undergo hydrolysis in the presence of the ammonia which has been liberated; for example:

$$[Ti(NH_3)_6]^{3+} + 6H_2O \rightarrow [Ti(H_2O)_6]^{3+} + 6NH_3$$

and

$$[Ti(H_2O)_6]^{3+} + 3H_2O \rightarrow [Ti(OH)_3(H_2O)_3] + 3H_3O^+$$

(removed by the ammonia produced in the first reaction)

Having considered ammonia as our prime example of a neutral ligand, some mention of other uncharged donors should now be made. Other monodentate nitrogen donors behave rather similarly to ammonia in that they often cause precipitation of hydroxides in aqueous solution. Metals that form ammines, however, form similar complexes with, for example, methylamine, CH_3NH_2, and pyridine, C_5H_5N, except that the position of the equilibrium may not be exactly the same as that found for ammonia. With pyridine particularly, the donor strength of the anion plays a part in determining the nature of the product formed. Nickel(II) nitrate solutions yield $[Ni(py)_6](NO_3)_2$ which can be isolated; the chloride, however, yields the blue $NiCl_2(py)_4$ in which the chloride ions have entered the coordination sphere of the nickel. Monodentate oxygen and sulphur donors similarly form complexes by substitution reactions. Urea, for example, forms $Fe[OC(NH_2)_2]_6^{3+}$ and thiourea $Ni(SC(NH_2)_2]_6^{2+}$. Ethanol and ether do not usually effect a great replacement of coordinated water molecules but complexes such as $[Ni(C_2H_5OH)_6](NO_3)_2$ can be prepared using ethanolic solutions in the complete absence of water. In absolute ethanol, complexes undergo deprotonation just as in the case of water, and alkoxides (rather than hydroxides) are formed, especially in the presence of added base.

Bidentate ligands, especially those capable of ring formation with the metal, form more stable complexes than those of monodentate ligands, having the same donor atom, by the chelate effect (table 8). In copper(II) chemistry, only four of the water molecules are replaced by, for example, ethane-1,2-diamine (as was the case with ammonia); complete substitution occurs, however, with those metals which can form hexammines in aqueous solution. Examples of bidentate ligands reacting in aqueous solution are given in the equations below (abbreviations for ligands are given in Chapter 2).

$$[Cu(H_2O)_6]^{2+} + 2en \rightarrow [Cu(en)_2(H_2O)_2]^{2+} + 4H_2O$$
$$[Ni(H_2O)_6]^{2+} + 3en \rightarrow [Ni(en)_3]^{2+} + 6H_2O$$
$$[Fe(H_2O)_6]^{3+} + 3bipy \rightarrow [Fe(bipy)_3]^{3+} + 6H_2O$$
$$[Fe(H_2O)_6]^{2+} + 3phen \rightarrow [Fe(phen)_3]^{2+} + 6H_2O$$

The last complex, as tris(o-phenanthroline)iron(II) sulphate, is widely used in volumetric analysis as 'ferroin'; the colour change with strong oxidizing agents in aqueous solution is quite striking:

$$\underset{\text{deep red}}{[Fe(phen)_3]^{2+}} \rightarrow \underset{\text{pale blue}}{[Fe(phen)_3]^{3+}} + e^-$$

Substitution by anionic ligands

The reactions between hexaquo ions and anionic ligands often result in the complete replacement of water molecules with the formation of anionic complexes. This is most commonly true of the ligands OH^-, F^-, Cl^-, Br^-, I^-, and CN^-. We have considered hydroxo complexes in the discussion of amphoterism on page 30 and we shall consider mainly the halide and cyanide ligands here. The principal complex species formed between metal ions in aqueous solution and an excess of added chloride or cyanide ions are listed in table 10. These species are those normally formed when any simple salt of the metal in question is treated with either hydrochloric acid or potassium cyanide. Other complex species, in addition to those listed in the table, are formed—for example, monosubstituted species such as $[Fe(H_2O)_5Cl]^{2+}$—but for the sake of simplicity we shall consider here only those principally occurring species found when an excess of the ligand is present. One difference from the substitution reactions by neutral ligands is that a change of coordination number and stereochemistry often occurs in substitution reactions by anions. A rather general reaction with chloride ions is the formation of tetrahedral complexes:

$$\underset{\text{octahedral}}{[M(H_2O)_6]^{n+}} + 4Cl^- \rightleftharpoons \underset{\text{tetrahedral}}{[MCl_4]^{(4-n)-}} + 6H_2O$$

The equilibrium position of this reaction is forced to the right-hand side if a high concentration of chloride ions is present (e.g. by use of concentrated hydrochloric acid) or if the concentration of water, relative to that of chloride ion, is lowered (e.g. by use of ethanolic solutions). For Fe^{2+}, Fe^{3+}, Co^{2+}, and Cu^{2+}, the tetrachloro species are formed in aqueous solution without difficulty by the addition of a soluble chloride.

Metal ion	**Chloro complex**	**Cyano complex**
Al^{3+}	$[AlCl_4]^-$	(hydrolysis)
Mn^{2+}	$[MnCl_4]^{2-}$	$[Mn(CN)_6]^{4-}$
Fe^{2+}	$[FeCl_4]^{2-}$	$[Fe(CN)_6]^{4-}$
Fe^{3+}	$[FeCl_4]^-$	$[Fe(CN)_6]^{3-}$
Co^{2+}	$[CoCl_4]^{2-}$	$[Co(CN)_5(H_2O)]^{3-}$
Co^{3+}	(reduction)	$[Co(CN)_6]^{3-}$
Ni^{2+}	$[NiCl_4]^{2-}$	$[Ni(CN)_4]^{2-}$
Cu^+	$[CuCl_2]^-$, $[CuCl_3]^{2-}$, $[CuCl_4]^{3-}$	$[Cu(CN)_2]^-$, $[Cu(CN)_4]^{3-}$
Cu^{2+}	$[CuCl_4]^{2-}$	$[Cu(CN)_4]^{2-}$ (with reduction)
Ag^+	$[AgCl_2]^-$	$[Ag(CN)_2]^-$
Hg^{2+}	$[HgCl_4]^{2-}$	$[Hg(CN)_4]^{2-}$

Table 10 Anionic complexes of chloride and cyanide ion formed in aqueous or ethanolic solution.

With Mn^{2+} and Ni^{2+}, the chloro complexes are not stable in aqueous solution and so ethanolic solutions of soluble chlorides are used to convert manganese(II) and nickel(II) salts to their tetrachloro complexes. In order to isolate any of these complexes, it is in any case preferable to use ethanolic solutions from which the compounds crystallize more readily (they are very soluble in water). Thus $(Et_4N)_2[FeCl_4]$ crystallizes as cream-coloured needles when solutions of tetraethylammonium chloride and hydrated iron(II) chloride in ethanol are mixed.

Perhaps the most interesting of these reactions is that of cobalt(II). Any soluble cobalt(II) salt, such as the chloride, sulphate, or nitrate, when treated with concentrated hydrochloric acid, undergoes the remarkable colour change from pink to blue:

$$[Co(H_2O)_6]^{2+} + 4Cl^- \rightleftharpoons [CoCl_4]^{2-} + 6H_2O$$

pink blue

The majority of cobalt(II) compounds are in fact either pink or blue; the occurrence of these two types of cobalt(II) compound was the object of chemists' curiosity for many years. The realization that the colour of the compounds is associated with their stereochemistry is of relatively recent origin. While there are notable exceptions, the majority of the pale-pink or orange cobalt(II) compounds have an octahedral structure, while the deep-blue compounds contain tetrahedrally coordinated cobalt(II). The structural change occurring in the change from pink $[Co(H_2O)_6]^{2+}$ to blue $[CoCl_4]^{2-}$ is thus:

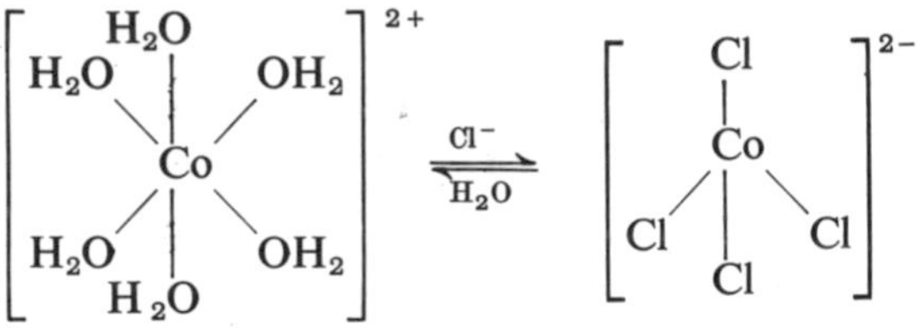

$$[Co(H_2O)_6]^{2+} \underset{H_2O}{\overset{Cl^-}{\rightleftharpoons}} [CoCl_4]^{2-}$$

This colour change is used in the drying agent, silica gel. The silica is treated with a cobalt(II) salt, as indicator, which is blue when anhydrous and becomes pink upon hydration. Thus, when the drying agent is blue it is still effective, but when it is saturated with water the colour changes to pink as the $[Co(H_2O)_6]^{2+}$ ion is formed. The gel can then be redried in an oven at 120 °C when the blue colour returns as the water is evolved. Unfortunately, in pale-blue cobalt(II) chloride the cobalt is octahedrally surrounded by chloride ions in the solid state, so that in this example the colour change is not due to the tetrahedral–octahedral interconversion.

Let us consider one further case, that of copper(II). Treatment of copper(II) sulphate solution with sodium chloride or concentrated hydrochloric acid brings about a colour change from blue through green to yellow. It is incorrect to say that copper(II) chloride is formed in the reaction and no satisfactory equation can be written if $CuSO_4$ is included as reactant. The blue colour of copper(II) sulphate solution is due to the $[Cu(H_2O)_6]^{2+}$ ion and the yellow colour to the $[CuCl_4]^{2-}$ ion:

$$[Cu(H_2O)_6]^{2+} + 4Cl^- \rightleftharpoons [CuCl_4]^{2-} + 6H_2O$$

blue yellow

The green colour observed at intermediate concentrations results from partial substitution by chloride ions, that is the blue and yellow ions together give a green coloration.

Why should the tetrahedral species, $[MCl_4]^{2-}$, be formed in preference to the octahedral species, $[MCl_6]^{4-}$? This question is a very difficult one and many factors are involved. Many metals do in fact retain the octahedral configuration. The elements in group IV form $[SiCl_6]^{2-}$, $[GeCl_6]^{2-}$, $[SnCl_6]^{2-}$, and $[PbCl_6]^{2-}$, for example, and hexaquo ions of Ti^{III}, V^{III}, and Cr^{III} give octahedral chloro species such as $[TiCl_5(H_2O)]^{2-}$ and $[CrCl_6]^{3-}$. The tetrahedral configuration is preferred for some ions, and the relatively small difference between the crystal field stabilization energies,* for the octahedral and tetrahedral configurations for these ions, is at least partly responsible for this. In a simple way, the formation of tetrahedral complexes with anions, when octahedral complexes occur with neutral ligands, can be rationalized in terms of size and charge of anions. Chloride ions are larger than water molecules and are negatively charged. Repulsion between the Cl^- ions will obviously be less in, say, $[CoCl_4]^{2-}$, where they are tetrahedrally separated, than in octahedral $[CoCl_6]^{4-}$, and so the tetrahedral structure would be expected to be somewhat more stable. This increase in stability has to be compared with the decreased stability as compared to the octahedral structure arising from crystal field stabilization energy effects. When the anion is small, the repulsion between anions becomes less important and octahedral complexes become more common. Aluminium, for example, forms $[AlF_6]^{3-}$ (as in cryolite) with fluoride ions but $[AlCl_4]^-$ with chloride ions. Similarly, if the anion is more polarizable, its charge cloud is more easily deformed and again octahedral coordination may result. Cyanide ions thus give predominantly octahedral complexes.

Those metals (notably Cu^I and Ag^I), which have a preference for linear two-coordination, form complexes of this shape irrespective of whether anions or neutral molecules are the ligands. Silver chloride, for example, dissolves in a number of aqueous reagents and in each case soluble two-coordinate complexes are formed:

$$AgCl \xrightarrow{NH_3(aq)} [Ag(NH_3)_2]^+ \qquad AgCl \xrightarrow{\text{conc. HCl}} [AgCl_2]^- \qquad AgCl \xrightarrow{KCN(aq)} [Ag(CN)_2]^-$$

Bromide and iodide complexes usually have the same structures as the chloride analogues. The reactions of these ions with transition metal aquo ions differs, however, in that they are more readily oxidized to the halogen by the metal. Bromide and iodide ions thus frequently act as reducing agents to metal ions which are in their highest oxidation states. Thus copper(II) is reduced to copper(I) by iodide ions in aqueous

*This is too complex a subject to be dealt with here. Readers who are interested should consult more advanced books such as *Advanced Inorganic Chemistry* by F. A. Cotton and G. Wilkinson, third edition, Interscience, 1972, or *Complexes and First Row Transition Elements* by David Nicholls, Macmillan, 1974.

solution, as it is also by cyanide ions:

$$[Cu(H_2O)_6]^{2+} \underset{H_2O}{\overset{I^-}{\rightleftharpoons}} [CuI_4]^{2-} \rightarrow CuI + 2I^- + \tfrac{1}{2}I_2$$

$$CN^- \downarrow \uparrow H_2O$$

$$[Cu(CN)_4]^{2-} \rightarrow CuCN + 2CN^- + \tfrac{1}{2}(CN)_2$$

$$CN^- \downarrow \uparrow H_2O$$

$$[Cu(CN)_4)^{3-}$$

Copper(I) iodide and cyanide are precipitated in these reactions but the cyanide is soluble in an excess of potassium cyanide solution. Vanadium, which shows oxidation states of +5, +4, +3, and +2 in aqueous media, illustrates the gradation in the reducing power of the halide ions. Thus vanadate(V) ions, $[VO_4]^{3-}$, are not reduced by hydrofluoric acid, but are reduced to vanadium(IV) species (such as $[VOCl_4]^{2-}$) by hydrochloric acid, and to vanadium(III) species (such as $[VBr_2(H_2O)_4]^+$) by hydrobromic acid. The mercury(II) ion is not reduced by iodide ions but, instead, red HgI_2 is precipitated, which can be dissolved in an excess of potassium iodide to form yellow $K_2[HgI_4]$. This compound has long been used, in alkaline solution, as a qualitative testing reagent for ammonia – Nessler's reagent. The reagent turns brown in the presence of a trace of ammonia; the reaction occurring is believed to be:

$$2[HgI_4]^{2-} + NH_3 + 3OH^- \rightarrow [Hg_2O(NH_2)]I + 2H_2O + 7I^-$$

The hexacyano complexes in table 10 have the octahedral structure. Nickel(II) salts react with aqueous potassium cyanide to give, initially, a green precipitate of nickel(II) cyanide; this is soluble in an excess of potassium cyanide resulting in an orange solution from which the yellow $K_2Ni(CN)_4 \cdot H_2O$ can be crystallized. In this salt (which can be dehydrated at 100 °C) the $[Ni(CN)_4]^{2-}$ ion has the square planar structure:

$$\left[\begin{array}{ccc} NC & & CN \\ & Ni & \\ NC & & CN \end{array}\right]^{2-}$$

Nickel also shows this tendency toward forming square complexes in its reaction with butanedione dioxime. This ligand is used in the gravimetric estimation of nickel because, with nickel ions in slightly alkaline solution, it gives a red precipitate of the very insoluble bis(butanedione dioximato)nickel(II):

$$[Ni(H_2O)_6]^{2+} + 2\begin{array}{c}\mathrm{MeC{=}NOH}\\ |\\ \mathrm{MeC{=}NOH}\end{array} \rightarrow \begin{array}{ccccc} & & \mathrm{O\cdots H{-}O} & & \\ \mathrm{MeC{=}N} & / & & \backslash & \mathrm{N{=}CMe}\\ | & \backslash & \mathrm{Ni} & / & |\\ | & / & & \backslash & |\\ \mathrm{MeC{=}N} & \backslash & & / & \mathrm{N{=}CMe}\\ & & \mathrm{O{-}H\cdots O} & & \end{array}$$

$$+2H^+ + 6H_2O$$

In this complex the four nitrogen atoms from the chelate are arranged in a square plane with the nickel ion in the centre. The dashed lines between the oxygen and hydrogen atoms represent hydrogen bonds.

The cobalt(II)–cyanide complex is unusual. The olive-green solution of cobalt(II) cyanide in potassium cyanide solution is rapidly oxidized by air; it is not known for certain whether it contains $[Co(CN)_5]^{3-}$ or $[Co(CN)_5(H_2O)]^{3-}$ ions. The addition of ethanol to the solution results in the precipitation of violet $K_6(Co_2(CN)_{10}]\cdot 4H_2O$ which is believed to possess a cobalt–cobalt bond in the anion, that is $[(NC)_5Co{-}Co(CN)_5]^{6-}$.

One or two other anionic ligands are worthy of mention. The thiocyanate ion forms a tetrahedral complex with cobalt(II) ions, that is $[Co(NCS)_4]^{2-}$, which can be readily precipitated as the blue, crystalline mercury salt, $Hg[Co(NCS)_4]$, by the addition of mercury(II) ions. Note that, as this complex contains nitrogen bonded to cobalt, it should be called an 'isothiocyanate' complex to distinguish it from sulphur-bonded thiocyanate complexes. It is used as a calibrant for magnetic susceptibility measurements. The reaction of thiocyanate ions with iron(III) is also of some practical importance. The blood-red colour developed in such solutions is due to the $[Fe(SCN)(H_2O)_5]^{2+}$ ion:

$$[Fe(H_2O)_6]^{3+} + SCN^- \rightleftharpoons [Fe(SCN)(H_2O)_5]^{2+} + H_2O$$

The nitrite ion forms the well-known 'cobaltinitrite' ion with cobalt(III). Sodium hexanitrocobaltate(III) is prepared by passing air through a mixture of cobalt(II) nitrate and sodium nitrite in the presence of ethanoic acid:

$$2[Co(H_2O)_6]^{2+} + 12NO_2^- + 2CH_3COOH + \tfrac{1}{2}O_2 \rightarrow 2[Co(NO_2)_6]^{3-} + 2CH_3COO^- + 13H_2O$$

It has been used as a qualitative reagent for potassium ions because of the insolubility of potassium hexanitrocobaltate(III).

Structure and Bonding 5

Coordination numbers and stereochemistry

In Chapter 2 we considered what types of species combine together to form complexes and in Chapter 4 we saw many examples of such complexes in chemical reactions. We must now consider the various coordination numbers which are known and in what shapes the ligands arrange themselves spatially around the metal ion. Werner found that certain metals have preferred coordination numbers (e.g. cobalt(III) six, platinum(II) four) and that the ligands arrange themselves around the metal in a preferred and definite shape. It would be very convenient if we could classify all the metals under preferred coordination numbers but unfortunately today most metallic ions are known to form complexes of several different coordination numbers and stereochemistries. It is still useful, however, to remember that cobalt(III) and chromium(III) form largely octahedral complexes, that beryllium and boron form largely tetrahedral complexes, that platinum(II) and palladium(II) form mostly square complexes, and that silver(I) forms mainly linear complexes. The structures of complexes are determined in the solid state by X-ray crystallography; by this technique the positions of the various atoms are directly located and bond lengths and angles estimated. In solution the structures of complexes are determined by less direct methods such as ultraviolet, visible, and infrared spectroscopy.

All coordination numbers from two to twelve are known in complex compounds. For the main group elements and elements in the first transition series, coordination numbers four and six are by far the most common, and we shall confine ourselves largely to these here. Coordination number two will be considered in view of its importance in copper(I) and silver(I) chemistry.

The shape of complex compounds formed by the main group elements can be predicted using electron-pair repulsion theory. Using this theory we assume that the outermost, or valency, electrons surrounding the central atom exist in localized pairs and that these electron pairs repel one another and so occupy positions as far apart as possible. Thus with two pairs of valency electrons in a complex, these will be at an angle of 180° to each other and the shape will be linear. With four electron pairs, the preferred shape is tetrahedral with a bond angle of 109° 28′ (the bond angles of electron pairs in the square planar structure are only 90°). In six-coordination the six electron pairs take up the octahedral distribution, all the bond angles being 90°. The shapes of transition metal complexes cannot always be understood on such a simple theory. Ions of configuration d^0, d^5, or d^{10}, that is those having empty, exactly half-filled, or completely filled d shells, do behave as expected from electron-pair repulsion theory because they have a symmetrical distribution of d electron density. Ions of other d electron configurations do not always follow these simple ideas and more advanced theories must be used to comprehend the shapes of their complexes. With coordination numbers two and six, however, we still find linear and octahedral coordination, respectively, predominating. Coordination number two occurs with Cu^I, Ag^I, Au^I, and Hg^{II}; for these elements it is a characteristic feature of their chemistry but for other elements it is exceptionally rare. Coordination number six occurs very commonly throughout the transition elements as well as among the main group elements (with the exception of the second-row elements Li to Ne which form tetrahedral complexes).

Coordination number four occurs largely with the tetrahedral shape. Main group elements possessing lone-pair electrons in addition to the bonding pairs of electrons may, however, be forced to take up the square planar shape as in, for example, XeF_4 and ICl_4^-:

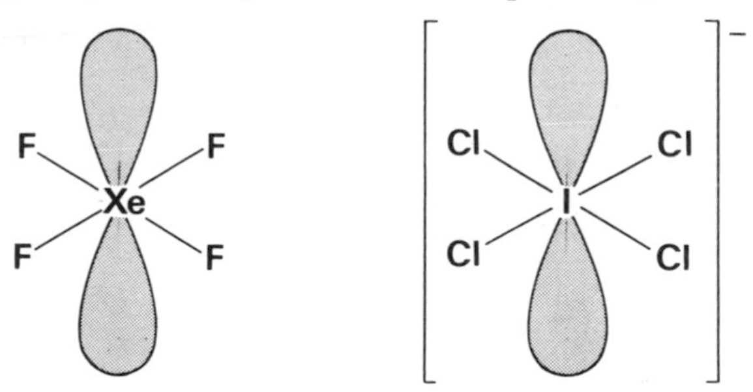

The shapes of these compounds are thus based upon the octahedron which is the shape expected in repulsion theory for six electron pairs. Among the transition elements the tetrahedral shape is still the most common for four-coordination. The square planar configuration is, however, a characteristic feature of the chemistry of platinum(II) and palladium(II) and occurs also with rhodium(I), iridium(I), gold(III), and nickel(II). Apart from these elements it occurs only very rarely. Examples of these stereochemistries are given in table 11.

Two-coordination	**Four-coordination**		**Six-coordination**
Linear	**Tetrahedral**	**Square**	**Octahedral**
$[Ag(CN)_2]^-$	$[FeCl_4]^{2-}$	$[PdCl_4]^{2-}$	$[Al(H_2O)_6]^{3+}$
$[Cu(NH_3)_2]^+$	$[Cd(NH_3)_4]^{2+}$	$[Pt(NH_3)_4]^{2+}$	$[Co(NH_3)_6]^{3+}$
$(Ph_3P)Au(C{\equiv}CPh)$	$CoCl_2(py)_2$	$[Ni(CN)_4]^{2-}$	$[TiCl_6]^{2-}$
$HgCl_2$	$Ni(CO)_4$	$[AuCl_4]^-$	SF_6

Table 11 Some examples of two-, four-, and six-coordination.

Isomerism

In Chapter 1 we mentioned briefly how the existence of isomers was important in enabling Werner to deduce the stereochemistry of six-coordination. Nowadays, many types of isomerism are known for metal complexes; we shall concern ourselves here with the two more important types, namely geometrical and optical isomerism.

Geometrical isomerism

This type of isomerism is otherwise known as *cis–trans* isomerism. Classic examples occur in the square platinum(II) complexes. The compound $PtCl_2(NH_3)_2$ for example is known in two different forms; these are called *cis*- or *trans*-$PtCl_2(NH_3)_2$ depending upon the relative positions of the atoms:

```
Cl       NH3          Cl       NH3
  \     /               \     /
    Pt                    Pt
  /     \               /     \
Cl       NH3        H3N        Cl
    cis                  trans
```

Notice that no isomers would be found if this complex had a tetrahedral structure since the Cl—Pt—Cl angle can only be 109° and the chlorine atoms must be adjacent. The following hypothetical structures for this compound are thus identical, merely being viewed from a different angle:

```
Cl       NH3          Cl       NH3
  \     /               \     /
    Pt                    Pt
  /     \               /     \
Cl       NH3        H3N        Cl
```

Octahedral complexes of the general type, MA_4B_2, similarly exhibit *cis–trans* isomerism; for example, the dichlorotetramminecobalt(III) cation:

```
[        Cl         ]+     [        Cl         ]+
[  Cl    |    NH3   ]      [  H3N   |    NH3   ]
[     \  |  /       ]      [     \  |  /       ]
[       Co          ]      [       Co          ]
[     /  |  \       ]      [     /  |  \       ]
[  H3N   |    NH3   ]      [  H3N   |    NH3   ]
[       NH3         ]      [        Cl         ]
        cis                        trans
```

Both such isomers can frequently be isolated but even when they cannot the nomenclature *cis* and *trans* is useful to describe whichever form has been isolated.

Optical isomerism

As with carbon compounds, in order for a complex to be optically active, that is capable of rotating the plane of polarized light, it must be asymmetric. It must have no plane of symmetry and the structure and its mirror image must be different, that is non-superimposable. A useful analogy is a pair of hands. Your left hand has no plane of symmetry and its mirror image, your right hand, is different—it cannot be superimposed on the left hand because of the different orientations of the thumb and fingers. In inorganic chemistry the largest class of optically active compounds is octahedral complexes containing two or three bidentate ligands—$M(L—L)_2X_2$ and $M(L—L)_3$. These ligands almost invariably contain carbon. Werner, endeavouring to prove that their activity was not necessarily due to the carbon present, succeeded in resolving the complex

$$\left[Co\left(\begin{matrix}H\\O\\ \\O\\H\end{matrix}Co(NH_3)_4\right)_3\right]Br_6$$

Simpler species than this capable of resolution include the tris(ethane-dioato) chromate(III) anion. The optical isomers can be illustrated as follows where for simplicity O—O represents $[O_2C—CO_2]^{2-}$:

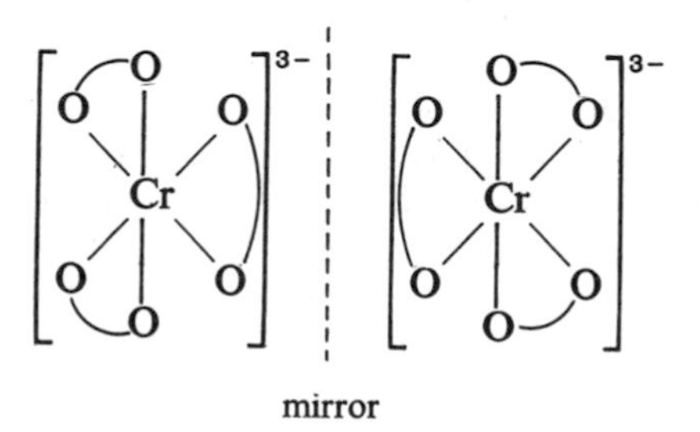

(You may find it necessary to build models of these anions to prove to yourself that they are not superimposable.)

The fact that optical isomers exist for these compounds is good evidence in favour of the octahedral structure. Neither the trigonal prismatic nor the hexagonal planar (alternative possibilities for six-coordination) structures would show optical activity. Racemization in inorganic complexes occurs very rapidly, however, so it is not always possible to separate such isomers. For complexes of the type $M(L—L)_2X_2$, optical activity shows the existence of the *cis*- structure since the *trans*-form has a plane of symmetry. The dichlorobis(ethane-1,2-diamine)-

cobalt(III) cation in $[Coen_2Cl_2]Cl$ thus exists in three forms, two optically active *cis* forms and an inactive *trans* form:

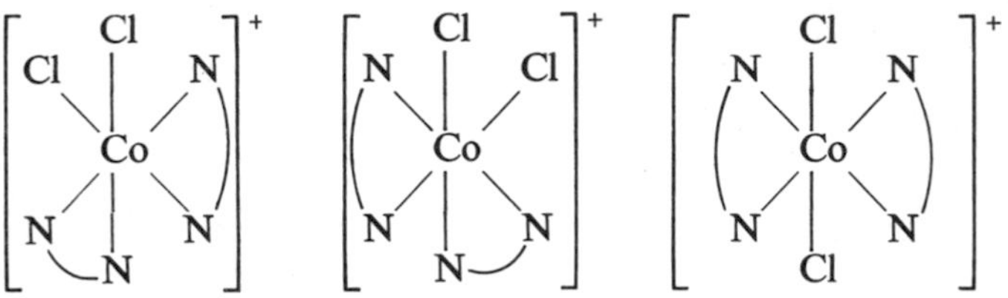

cis forms, active　　　　*trans*, inactive

(N—N represents $H_2NCH_2CH_2NH_2$.)

Bonding

It will not be possible here to give a complete description of our present-day understanding of the bonding in complex compounds. The subject is a very complicated one and involves a knowledge of the symmetry properties and energies of atomic and molecular orbitals. It will be necessary here for the author to assume that the student has had some contact with orbitals but only in a very elementary way.

We saw in Chapter 2 that complexes contain coordinate bonds formed by the donation of electron pairs from the ligands to the central atom or ion. This simple idea of Lewis and Sidgwick was developed by Pauling into the *valence bond theory* of the bonding in complexes. Whilst this theory finds great use in organic chemistry and in the chemistry of the main group s- and p-block elements it is now largely unsuitable for transition metal complexes. It has been replaced by a theory known as the *ligand field theory*. There are two extreme approaches that may be taken to this theory. In one, the ligands are regarded as point charges surrounding the central positively charged ion, and the electrostatic effect of these charges upon the energies of the metal d orbitals is considered. This is the *crystal field theory*. The alternative approach is to consider the effect of covalent bonding, occurring through the overlap of orbitals on the metal with those on the ligand, upon the energies of the d orbitals. This is *molecular orbital theory* applied to the d orbitals in complexes. Both crystal field and molecular orbital theories lead to the conclusion that in a complex compound the energies of the d orbitals are not all the same (as they would be in a free metal ion). We shall now use the molecular orbital approach to illustrate this and then illustrate the consequences of 'd orbital splitting'. We shall deal with the octahedral structure in detail because this is the most commonly occurring; similar arguments can be applied to other stereochemistries.

The ligand field theory

Covalent bonds are regarded as being formed by the overlap of orbitals on the metal ion and on the ligands. In order for this overlap to be significant and hence for the bonding to be effective, two criteria are required of the orbitals. Firstly, they must be of the same symmetry (shape) so that they do in fact overlap when a three-dimensional picture is considered. Secondly, they must be of similar energies. Before we can see if metal ion d orbitals can effectively overlap with ligand orbitals we need to know the shapes of the d orbitals. The approximate electron density distribution in the five d orbitals is indicated in figure 3. These orbitals are balloon-shaped—three-dimensional—and not flat as they appear in the figure. The first two, that is the d_{z^2} and the $d_{x^2-y^2}$ orbitals, lie along the directions of the cartesian axes. These two orbitals are at right angles to each other so that together they possess lobes pointing to the six corners of an octahedron. The other three orbitals are identical to each other except in their spatial orientation. The d_{xy} orbital lies in the *xy* plane with the lobes pointing between the *x* and *y* axes. Similarly, the d_{xz} and d_{yz} orbitals have their lobes pointing between the *x* and *z*, and the *y* and *z* axes respectively.

The orbitals on the ligand ions or molecules may be sp^3 hybrid orbitals or simple p orbitals. In an octahedral complex, six ligands, each

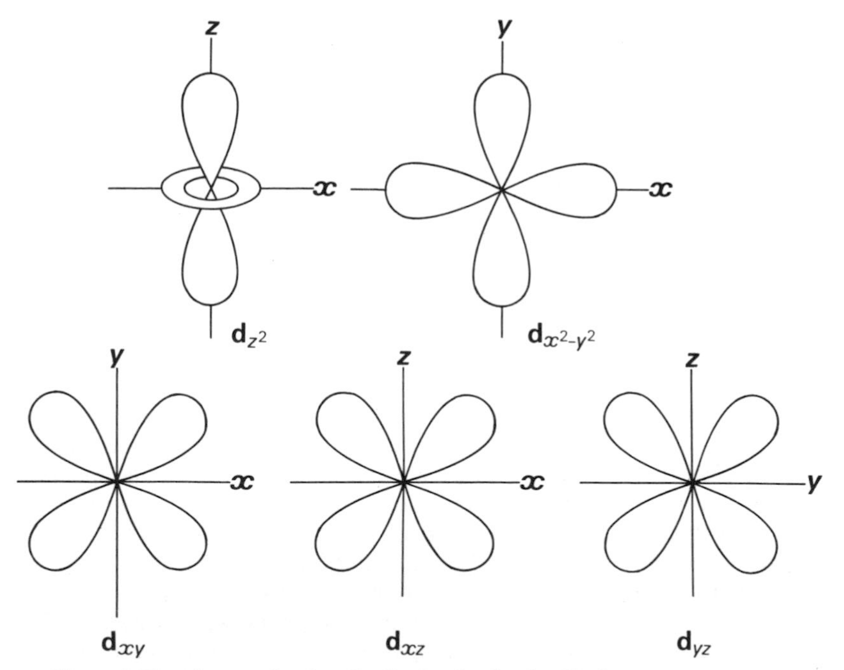

Figure 3 The electron density distribution in the d orbitals.

providing an orbital for bonding, surround the metal ion so that all the ligand–metal–ligand bond angles are 90°. These ligand orbitals have the right shape for overlap with the d_{z^2} and $d_{x^2-y^2}$ orbitals on the metal as shown in figure 4 (the collar on the d_{z^2} orbital has been left out of the picture for clarity). With the ligands lying right along the axes then, their orbitals can overlap with the metal orbitals which lie along the x, y, and z axes. The ligand orbitals do not have the correct symmetry to overlap with the d orbitals which point between the axes, that is the d_{xz}, d_{yz}, and d_{xy} orbitals. Thus no bonds (strictly no σ-bonds) can be formed with these orbitals—we say they are 'non-bonding' in an octahedral σ-bonded complex.

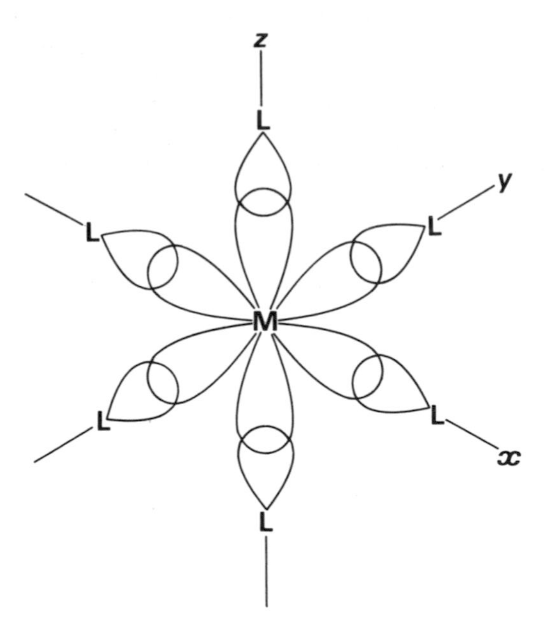

Figure 4 Overlap of ligand orbitals with metal d_{z2} and d_{x2-y2} orbitals in an octahedral ML_6 complex.

In molecular orbital theory we combine the orbitals on the metal and those on the ligand to give a set of molecular orbitals for the complex as a whole. When two atomic orbitals are combined, two molecular orbitals are produced. One of the molecular orbitals is lower in energy than the atomic orbitals—this is the *bonding* molecular orbital, and one is higher in energy—this is called the *antibonding* molecular orbital. Thus, when we consider the interaction of the ligand orbitals with the metal d orbitals we get the energy level diagram shown in figure 5. The energies of the non-bonding orbitals denoted by the letter n remain unchanged while the energies of the d_{z^2} and $d_{x^2-y^2}$ orbitals are raised to become largely the antibonding (denoted by *) $\sigma^*_{z^2}$ and $\sigma^*_{x^2-y^2}$. The ligand orbitals and the metal orbitals give the lowest energy bonding orbitals which have largely ligand character in the molecule and are denoted by σ_L. The letter σ merely denotes orbitals formed by end-on overlap (as opposed to side-on overlap) and distinguishes them from π orbitals with which we shall not be concerned.

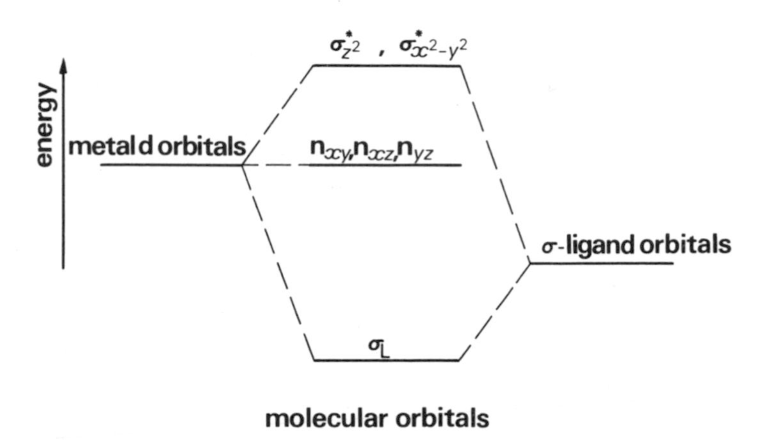

Figure 5 Energy levels arising from the interaction of metal d orbitals with σ-ligand orbitals.

The ligand orbitals also overlap with the s and p orbitals on the metal (these will be 4s and 4p for a transition metal of configuration $3d^n$) and produce further antibonding orbitals higher in energy than $\sigma^*_{z^2}$ and $\sigma^*_{x^2-y^2}$, and bonding orbitals lower in energy than those denoted by σ_L in figure 5. This gives us the complete molecular orbital picture which we do not need in ligand field theory.

The important result for ligand field theory then is that the d orbitals on the metal are split into two different energy sets when an octahedral complex is formed. The ligand field splitting diagram is usually drawn in the form shown in figure 6. The d orbitals all have the same energy in a free metal ion: we say they are degenerate. In an octahedral complex they are split into two sets. The set labelled t_{2g} correspond to the non-bonding orbitals labelled n in figure 5, and the set labelled e_g correspond to the antibonding orbitals in figure 5. When a complex is formed by donation of electron pairs from the ligand, these electrons occupy the low energy bonding molecular orbitals (σ_L) and need not concern us here. In ligand field theory we consider just the consequences of the splitting of the d orbitals. In figure 6 each horizontal line represents an orbital capable of containing up to two electrons. The electrons in these levels will be those contained in the free metallic ion (but not any from the ligand which as explained above are in lower energy orbitals).

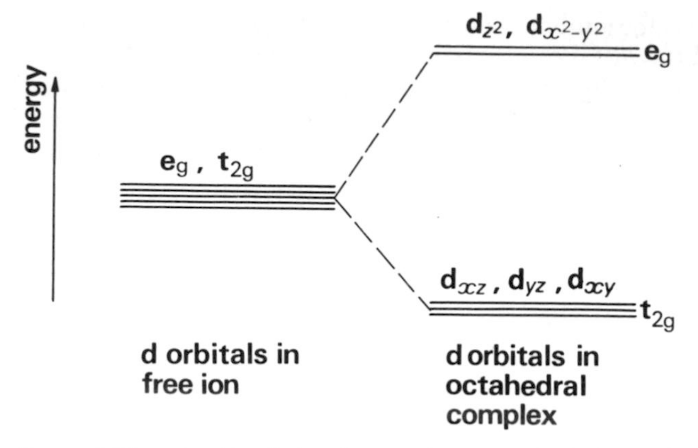

Figure 6 Ligand field splitting diagram for an octahedral complex.

In describing the structure of a complex we first determine the oxidation state of the metal ion and thus how many d electrons it contains. The simplest case to take is the hexaquotitanium(III) ion, $[Ti(H_2O)_6]^{3+}$. This pink–violet ion is readily obtained in the laboratory by dissolving titanium granules in hydrochloric acid and diluting the final solution somewhat to remove chloride ions from the coordination sphere. In this ion, the lone-pair electrons from sp^3 hybrid orbitals on the water molecules are donated into the low-lying bonding σ_L orbitals in the complex, and we are left with the d levels to fill with any electrons on the metal. Titanium has the ground state electron configuration of $[Ar]3d^24s^2$; in $[Ti(H_2O)_6]^{3+}$, titanium is in oxidation state +3. Titanium has thus lost three of its outermost electrons, and for Ti^{3+} the electron configuration becomes $[Ar]3d^1$. We thus have one electron to place in the d orbitals and we place it in the lower energy t_{2g} orbitals. Both the t_{2g} and the e_g sets of orbitals are degenerate so that it doesn't matter which of the t_{2g} levels we put the electron in. A diagrammatic description of the $[Ti(H_2O)_6]^{3+}$ ion using ligand field theory is thus:

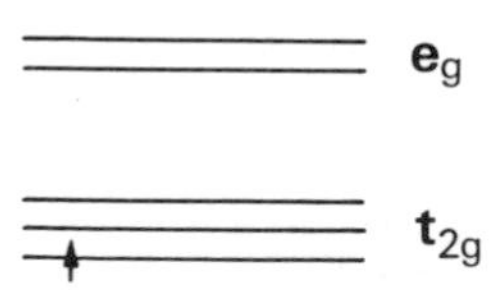

For ions of configuration d^2 or d^3 we can similarly feed in electrons to give the configurations t_{2g}^2 and t_{2g}^3 with two and three unpaired electrons respectively. For ions of configurations d^4, d^5, d^6, and d^7, however, there are two possible ways of feeding the electrons into the energy levels. These are illustrated on the next page for the d^6 configuration.

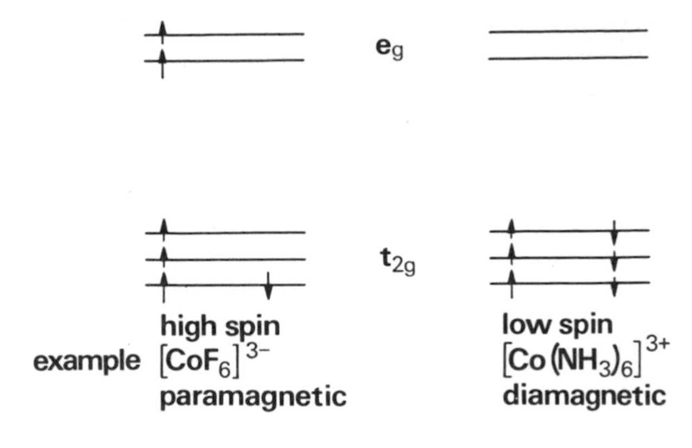

Ligand field theory thus explains the occurrence for these ions of two classes of octahedral complexes having different magnetic and spectroscopic properties (see below). Whether or not we get high-spin complexes depends upon the ligand. With strong field ligands (see later) the electrons reach a lower energy state by pairing up in the lower energy orbitals, while in weak ligand fields they reach a lower energy configuration by having the maximum number of unpaired electrons.

Magnetism in inorganic complexes

What does the diagram above tell us about the $[Ti(H_2O)_6]^{3+}$ ion? Firstly, it shows the presence of one unpaired electron. Now, compounds containing unpaired electrons exhibit the property of *paramagnetism*. Paramagnetic species are attracted toward a magnetic field while *diamagnetic* substances, containing only paired electrons, are weakly repelled by a magnetic field. Thus the $[Ti(H_2O)_6]^{3+}$ ion will be paramagnetic. The extent of this paramagnetism is related to the number of unpaired electrons. In the first transition series the magnetic moment of a metal ion is approximately given by the relationship known as the 'spin only' formula:

$$\mu = \sqrt{n(n+2)}$$

where μ denotes the magnetic moment (in Bohr magnetons*) and n denotes the number of unpaired electrons per metal ion.

For one unpaired electron, then, we expect $\mu = \sqrt{3} = 1.73$ BM and this is, indeed, the magnetic moment found (1.75–1.80) for the Ti^{3+} ion in $CsTi(SO_4)_2 \cdot 12H_2O$ which contains the $[Ti(H_2O)_6]^{3+}$ ion in the solid state. Similarly, for two unpaired electrons we should expect $\mu =$

*The Bohr magneton (abbreviated BM) is a unit which is used to measure magnetic moments. It is defined as 1 BM $= (eh)/(4\pi m_e)$ where e denotes the charge of the electron, h the Planck constant, and m_e the mass of the electron.

2.8 BM, for three unpaired electrons $\mu = 3.9$ BM, and so on. Since magnetic moments are easily measured in the laboratory they have played an important part in the testing of valency theories.

Spectroscopy and colour in complexes

The second observation that we might make concerning the ligand field splitting diagram for the $[Ti(H_2O)_6]^{3+}$ ion concerns spectroscopy. It should be possible to excite the electron from the t_{2g} level to the e_g level with visible or ultraviolet light. As the ion is coloured we know that absorption of light occurs in the visible region of the spectrum and that this absorption corresponds to the electron being excited from the t_{2g} to the e_g level:

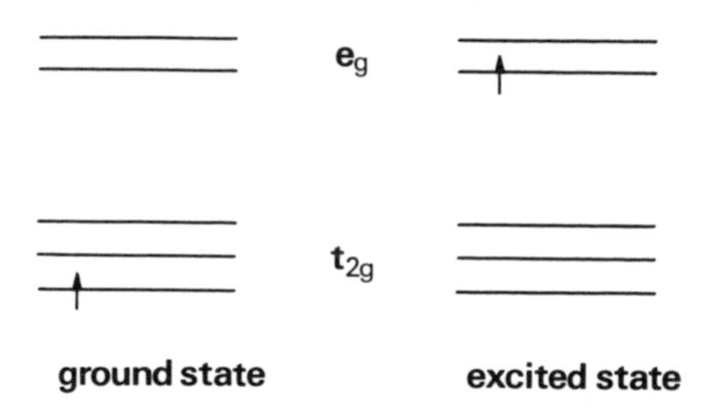

The visible absorption spectrum of the solution of titanium in dilute hydrochloric acid is shown in figure 7. This ion looks purple because it is absorbing orange, yellow, and green light but not blue light (around 25 000 cm^{-1}) or red light (below 15 000 cm^{-1}). Thus the light transmitted to the eye is red–blue or purple. The absorption band is rather broad and its maximum occurs at about 20 000 cm^{-1}. We say that the $t_{2g} \rightarrow e_g$ electronic transition occurs at 20 000 cm^{-1}. Now Planck's relationship

$$E = h\nu$$

shows that energy E and frequency ν are directly related by a constant h known as the Planck constant. By measuring the absorption frequency of the $[Ti(H_2O)_6]^{3+}$ ion we have thus measured experimentally the energy difference between the t_{2g} and the e_g orbitals. With the help of magnetism and electronic spectroscopy we are thus able to determine the number of unpaired electrons per metal ion (and hence the oxidation state of the metal) and the energy separation between the d orbitals.

The colour of transition metal complexes thus usually arises from the electronic transitions between the d orbitals. Species with an empty d shell, that is d^0 configuration, cannot normally show colour because there are no d electrons to excite. Similarly, species having a completely filled d shell, that is d^{10} configuration, do not show colour because there is no vacant level of sufficiently low energy for the electrons to be

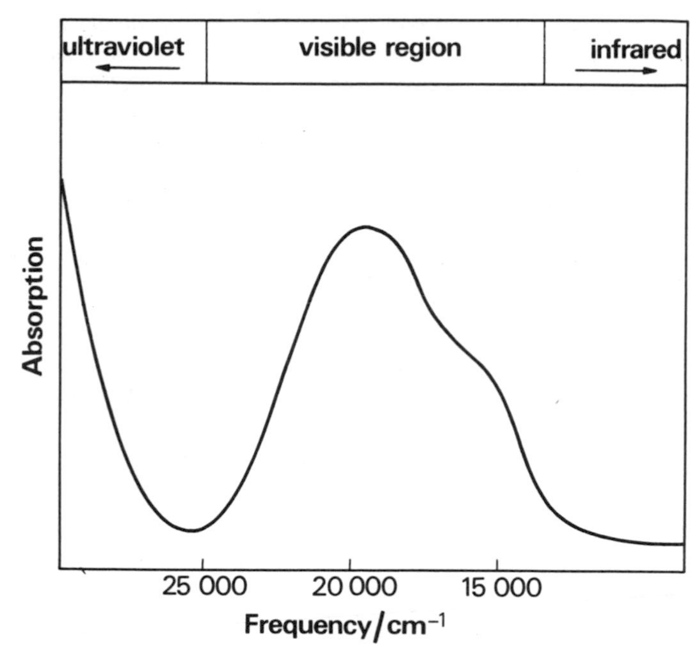

Figure 7 The visible absorption spectrum of the $[Ti(H_2O)_6]^{3+}$ ion.

excited into. The ligand field diagrams for octahedral d^0 and d^{10} ions are thus:

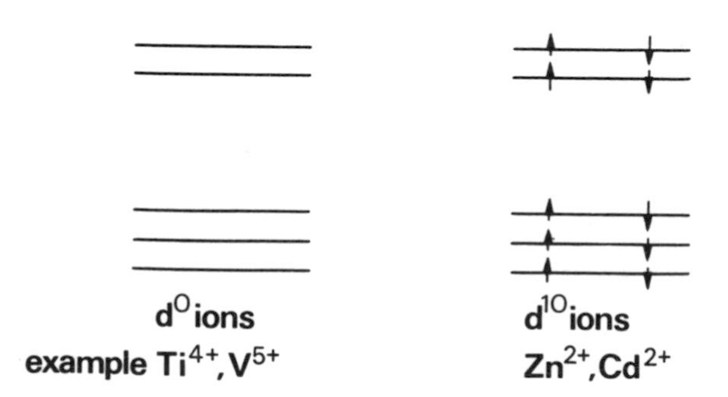

Thus, titanium(IV), vanadium(V), zinc(II), and cadmium(II) frequently form white (colourless) complexes; they give coloured compounds of course if the anion is coloured, for instance permanganate, or if colour arises from charge transfer. Similar arguments apply for other stereochemistries—the tetrahedral ligand field splitting diagram will be discussed shortly. Copper(I), $3d^{10}$, and silver(I), $4d^{10}$ will also be expected to form predominantly white complexes. We expect coloured complexes from transition metal ions having incomplete d electron shells.

The question which we should now ask ourselves is why do complexes of a given metal ion exhibit such a variation in colour? The reason is that the energy difference (Δ) between the two sets of d orbitals depends upon the ligand. Those ligands such as CN^- and CO which cause a large energy splitting are called strong field ligands while those which give the smallest energy splitting such as I^- are called weak field ligands. If we measure the visible spectra of many complexes we find that there is an order of ligand field strength which is true irrespective of the metal. The ligands arranged in order of ligand field strength give a series known as the *spectrochemical series*. This series for the more common ligands is:

$$I^- < Br^- < Cl^- < F^- \approx OH^- < H_2O < py$$
$$\approx NH_3 < en < bipy < CO < CN^-$$

weak field ligands   strong field ligands

How does this series help us to explain the colours of say copper(II) complexes? Copper(II) sulphate is white, the $[Cu(H_2O)_6]^{2+}$ ion is blue, the $[Cu(NH_3)_4(H_2O)_2]^{2+}$ ion is blue–violet, and the unstable $[Cu(CN)_4]^{2-}$ ion is colourless in solution. In anhydrous copper(II) sulphate there are only oxygen atoms from sulphate ions in the ligand field of the copper ion and in this weak field situation the electronic transition occurs in the lower energy region of the spectrum, that is the infrared. We cannot see this and so the compound looks white. In going from the blue aquo ion to the diaquotetrammine ion, the ligand field has moved up the spectrochemical series from $6H_2O$ to $4NH_3 + 2H_2O$ with the expected colour change. If we add cyanide ions to this solution the ligand field becomes $4CN^-$ in $[Cu(CN)_4]^{2-}$, that is the energy levels are separated even more and the energy necessary for the electronic transition has moved into the ultraviolet. The complex is still absorbing radiation but as it is not in the visible region we cannot see it. In this complex of course we no longer have octahedral stereochemistry so that our energy level diagram does not apply; other stereochemistries also exhibit energy level splittings in the d orbitals, however, so that the general ideas of this argument still apply. We also have reduction to copper(I) complexes occurring in this system, the extent of which will depend upon the conditions used; we have already seen that Cu^I, d^{10} complexes have no vacancy in the 3d shells and so are usually colourless (white in the solid state).

Distortion from octahedral symmetry—the Jahn–Teller effect

If we examine the spectrum of $[Ti(H_2O)_6]^{3+}$ (figure 7) in more detail we see that it is not composed of a single symmetrical band. Rather it has a bump or 'shoulder' on the low frequency side. The spectrum of an aqueous solution of copper(II) sulphate, that is $[Cu(H_2O)_6]^{2+}$, similarly shows a broad unsymmetrical band. Let us write down the ligand field splitting diagram for $[Cu(H_2O)_6]^{2+}$. Copper(II) has the outer electron configuration $3d^9$ so that we feed nine electrons into the d levels as follows (arrows pointing in different directions represent electrons having different spins):

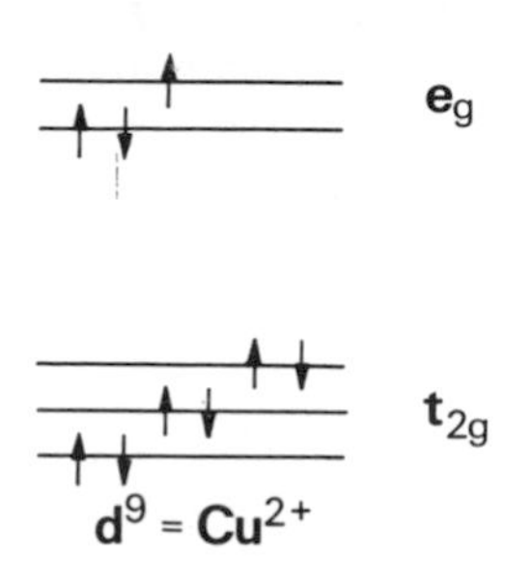

It is apparent from this diagram that we have two electrons in one of the e_g orbitals and only one electron in the other. Let us suppose that the d_{z^2} orbital has the two electrons and the $d_{x^2-y^2}$ the one electron. Now let us consider the interactions between the electron pair on each bonded ligand and the positive nucleus of the Cu^{2+} ion. The ligands in the *xy* plane will be less screened from the attractive forces of the Cu^{2+} ion than those along the *z* axis which have an extra electron. Consequently, the four ligands along the *xy* axes will approach closer to the metal ion than will those along the *z* axis. Alternatively, if we had chosen the $d_{x^2-y^2}$ orbital as the doubly filled orbital, then it would be the ligands along the *z* axis which approach closer to the nucleus. In either case the net result is that the octahedral structure distorts. This distortion is in these cases a tetragonal distortion and can be represented as shown in figure 8. The octahedral structure (*a*) is converted into the tetragonal structures (*b*) and (*c*) by compression or elongation along one of the axes (the *z* axis as drawn).

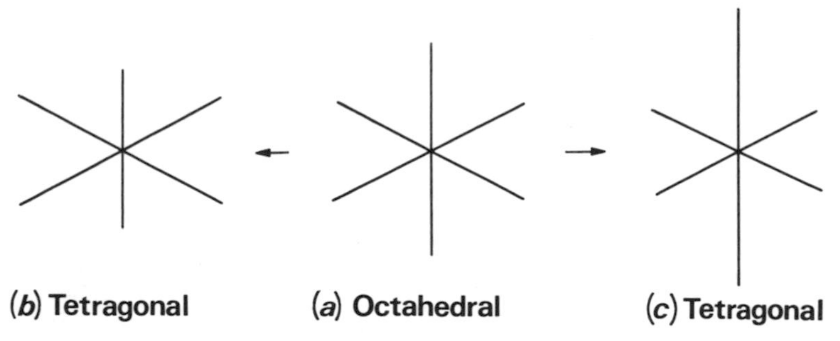

Figure 8 Tetragonal distortions of the octahedral structure.

Such distortions of perfect symmetries are common in inorganic complexes. The effect is known as the Jahn–Teller effect after the workers who first formulated the idea mathematically in 1937. Unfortunately the Jahn–Teller theorem does not tell us in which way a distortion will occur, merely that it will occur. With copper(II), the elongated tetragonal structure is the one commonly found. We saw in Chapter 3, figure 2, that in $CuSO_4 \cdot 5H_2O$ the Cu—O bonds to sulphate ions are longer than the Cu—O bonds to water molecules. The structure of the hexaquocopper(II) ion is thus tetragonal with two long bonds:

```
[        H2O            ]2+
[  H2O    |     OH2     ]
[      \  |  /          ]
[        Cu             ]
[      /  |  \          ]
[  H2O    |     OH2     ]
[        H2O            ]
```

The effect of such distortion on the ligand field splitting diagram is to remove the degeneracy of the e_g and t_{2g} levels. This is shown in figure 9. The electrons in the Cu^{2+} ion prefer energetically to be in the d_{z^2} orbital, that is further away from the electron density of the ligand water molecules, than in the $d_{x^2-y^2}$ orbitals. Thus the d_{z^2} orbital has a lower energy than the $d_{x^2-y^2}$ orbital. Similarly, the d_{xz}, d_{yz} pair having a component in the *z* axis are lower in energy than the d_{xy} orbital. Indeed the driving force for the distortion is that a lower energy, i.e. a more stable structure, is produced. In figure 9 it can be seen that the energy of

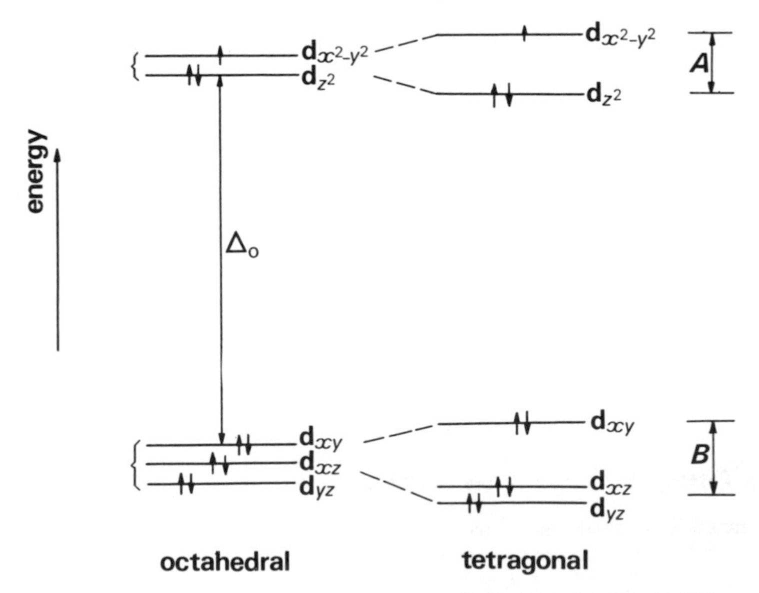

Figure 9 Ligand field diagram for a tetragonally elongated octahedral d^9 complex (bracketed levels are degenerate).

the tetragonal structure is less than that of the octahedral one by an amount corresponding to $\frac{1}{2}A$. We started in the octahedral structure with three electrons in the e_g level; we raised one in energy by $\frac{1}{2}A$ and lowered two in energy by $\frac{1}{2}A$. Hence the net result and a more stable structure. There is no net difference in energy in the t_{2g} electrons (two go up in energy by $\frac{2}{3}B$ and four come down in energy by $\frac{1}{3}B$). It should be pointed out that figure 9 is not drawn to scale; the splittings A and B are smaller relative to Δ_0 than is represented, and A is in fact larger than B.

The copper(II) ion is the classic example of the Jahn–Teller distortion. A smaller distortion occurs for octahedral complexes having no electrons in the e_g levels but having 1, 2, 4, or 5 electrons in the t_{2g} levels. In the excited state of the $[Ti(H_2O)_6]^{3+}$ ion, for example, we have the configuration $t_{2g}^0\ e_g^1$ so that this state undergoes the splitting similar to that in the Cu^{2+} case. The net result of these splittings is that instead of just one band being observed in the spectrum we see two or more bands close in energy because there are now two different energy levels into which the electron may be excited.

Tetrahedral complexes

So far we have considered only octahedral stereochemistry. The arguments which were used to establish the splitting of the d orbitals

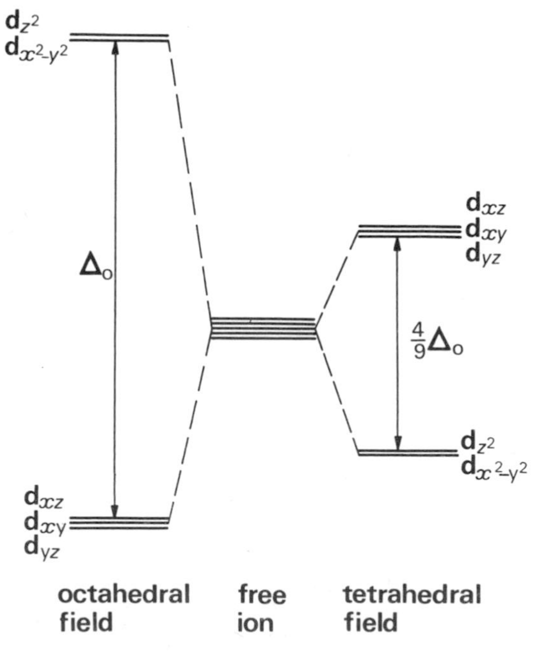

Figure 10 Ligand field splitting for octahedral and tetrahedral fields compared.

in octahedral complexes can be applied also to tetrahedral species. In tetrahedral complexes, however, the orbitals with the better symmetry for bonding with ligands are the d_{xy}, d_{xz}, d_{yz} set. Even these do not point directly at the ligands, however, so that the preference for these orbitals as opposed to the d_{z^2} and $d_{x^2-y^2}$ set is not so great relatively as it was for the e_g set in octahedral fields. The splitting of the d levels is therefore not so great here but is inverted as compared to the octahedral case. This is shown in figure 10. If we denote the energy separation of the d orbitals in an octahedral field as Δ_o then the corresponding tetrahedral separation is $\frac{4}{9}\Delta_o$. For tetrahedral complexes, we feed the available d electrons on the metal into the available levels as for octahedral complexes. Deductions from magnetic and spectral properties can then follow. Because of the lower energy separation between the d levels, nearly all tetrahedral complexes are of the high-spin type, that is electrons fill the levels so as to leave the maximum number of unpaired electron spins.

6 Some Practical Work and Revision Exercises

The exercises which follow are designed for laboratory use, and common chemicals are written in laboratory shorthand (e.g. dil. HCl) for simplicity. When performing the experiments, the student should make careful observations and record these in a laboratory notebook before answering the questions which follow. The exercises are intended to illustrate the theoretical chapters in this book and to show the reader how the understanding of inorganic reactions is simplified through the use of complex ions. Since answers to the questions are given at the end of this chapter, the reader may alternatively wish to use the exercises as revision tests or study problems. The answers given by the author have deliberately been kept to a minimum. These answers should serve to direct the reader to that section of the book which should be read again in the event of difficulty or if elaboration is required. The question numbers are placed in brackets at the end of each question and it is to these numbers that the answers correspond. The elements chosen for study are largely those for which the chemistry can be studied in aqueous solution. It is not practical to use boron trichloride or titanium(IV) chloride in a school laboratory, but the reactions of this type of covalent compound can be exemplified by tin(IV) chloride. The elements are not given in any special order and so may be used in any order by the reader. Having carried out experiments on two or three of the elements the reader will find that questions on the other elements become easier—this is one of the joys of systematizing inorganic chemistry.

Copper

Prepare an approximately 0.5 M solution of $CuSO_4 \cdot 5H_2O$ (about 10 g in 100 cm^3 of H_2O). What copper(II) ion is present in this solution?(1)

(i) Treat a little of this solution (1–2 cm^3) with conc. HCl, added a drop at a time to an excess. What copper species is now present in the solution? (2)

Criticize the equation:

$$CuSO_4 + 2HCl \rightarrow CuCl_2 + H_2SO_4$$

as an appropriate description of this experiment. (3)

Dilute the hydrochloric acid solution with five to ten times its own volume of water. Comment upon your observations. (4)

(ii) Treat 1–2 cm^3 of the $CuSO_4$ solution with conc. NH_3 solution added a drop at a time to an excess. What species are observed during this reaction? (5)

Why is $[Cu(NH_3)_6]^{2+}$ not formed? (6)

To a little of the solution obtained in excess of ammonia add potassium cyanide solution (DANGER—POISON) to an excess. What species is now present? (7)

Account for the colour change. (8)

(iii) Add an excess of solid EDTA (disodium salt of bis[di(carboxymethyl)amino]ethane) to a few cm^3 of the $CuSO_4$ solution and heat to dissolve. Now add ammonia solution a drop at a time. Explain your observations. (9)

(iv) To 5–10 cm^3 of the $CuSO_4$ solution add dil. NaOH solution to an excess. Filter off the precipitate (or centrifuge it), wash it two or three times with distilled water, and then dissolve it in dil. HCl. To this solution add $BaCl_2$ solution. Account for the formation of a white precipitate by tracing the course of the reactions. (10)

(v) Add about 2 g of Cu powder and 10 cm^3 of conc. HCl to 10 cm^3 of the $CuSO_4$ solution, and boil the mixture for several minutes. Decant or filter the solution from the excess of Cu into a large volume of water in a beaker. Filter off and wash this precipitate with distilled water and then add conc. NH_3 solution to it (while still in the filter paper). Explain the reactions occurring and account for the colours of the reaction products. (11)

Silver

In each of the following experiments (unless otherwise stated) use about 2 cm^3 of silver nitrate solution (about 0.05–0.1 M) in a test-tube or boiling-tube.

(i) Add dil. HCl to an excess (5 cm^3). What is the formula of the silver-containing product? (1)

(ii) Add about 1 cm^3 of dil. HCl followed by conc. NH_3 solution to an excess. What silver species is now present in the solution? (2)

(iii) Add conc. HCl slowly to an excess (10–20 cm^3). What species of silver is now present? (3)

(iv) Add dil. NaOH solution slowly to an excess. What is the silver product formed? (4)

(v) Add very dil. NH_3 solution a drop at a time to an excess. What silver species are formed in this reaction? (5)

(vi) Add a few drops of dilute potassium chromate solution. What is the formula of the precipitate? (6)
Add a little dil. HNO_3 to the mixture. Comment upon the use of potassium chromate as an indicator in volumetric titrations with silver nitrate solution. (7)

(vii) Add a few drops of NaCl solution followed by potassium cyanide solution (DANGER—POISON) to an excess. What silver species are formed in these reactions? (8)

(viii) Add dilute sodium thiosulphate solution slowly to an excess. What are the products formed? (9)
In what process is this reaction widely used? (10)

What general conclusion can you draw about the complexes of silver? (11)

Cobalt

In each of the following experiments use 1–2 cm^3 of a solution of cobalt(II) chloride, nitrate, or sulphate (about 1.0 M).

(i) Add a few drops of dil. NH_3 solution. Describe and give a formula for the precipitate. (1)

(ii) Add conc. NH_3 solution to an excess. Observe carefully, especially at the top of the test-tube where the solution is in contact with the air. To a separate portion of the cobalt solution add an excess of conc. NH_3 followed by a few drops of hydrogen peroxide (10 or 20 vol) solution. Explain the sequence of reactions occurring in these experiments. (2)

(iii) Add dil. NaOH solution a drop at a time to an excess. Observe and record carefully the colours of the precipitates seen. (3)

Heat the mixture gently (beware of bumping—use a boiling-tube if possible) until it is at the boiling point. What is happening to the precipitate now? (4)

(iv) Add potassium cyanide solution (DANGER—POISON) slowly to an excess. Observe the colour carefully (hold the tube against a white background) and then add a little hydrogen peroxide solution. Again observe the colour carefully. More cobalt(II) solution can be added to intensify the colour if necessary. What cobalt species are present in the original precipitate and the subsequent solutions? (5)

(v) Add conc. HCl slowly to an excess. Dilute the solution with water. Write an equation to describe the reactions occurring in this system. (6)

(vi) Add an excess of solid ammonium (or potassium) thiocyanate (NH_4SCN) until no more will dissolve and then add a few cm^3 of a concentrated solution of a mercury(II) salt (preferably $Hg(NO_3)_2$ which is more soluble than $HgCl_2$) (CARE—POISON). Allow the mixture to stand. Explain the chemistry involved in these reactions and give a structural formula for the product. (7)

(vii) Add an excess of solid potassium nitrite (KNO_2) and then acidify the mixture with dilute ethanoic acid. Explain the course of the reaction. (8)

Nickel

In each of the following experiments use 1–2 cm^3 of a solution of nickel(II) nitrate, sulphate, or chloride (about 1.0 M).

(i) Add very dil. NH_3 solution a drop at a time to an excess. Explain your observations. (1)

Perform these reactions in a slightly different way as follows. Prepare a saturated solution of the nickel salt (nitrate for solubility preference) in a test-tube. Add about 1 cm^3 of it to an excess of conc. NH_3 solution (10 cm^3) in a boiling-tube (add more ammonia if a clear solution is not obtained). Now add a few cm^3 of a saturated solution of ammonium bromide to the nickel–ammonia solution and allow the mixture to stand. What is the formula of the nickel-containing product? (2)

(ii) Add an excess of solid NH_4Cl followed by dil. NH_3 solution to an excess. Why does this reaction proceed differently from that in (i) above? (3)

(iii) Add dil. NaOH solution to an excess. Write an ionic equation for the reaction. (4)

(iv) Add potassium cyanide solution (DANGER—POISON) slowly to an excess. What nickel species are produced? (5)
What stereochemical shape does the species produced in an excess of the cyanide possess? (6)

(v) Add 1 or 2 cm^3 of butanedione dioxime (1 % in ethanol) solution followed by a few drops of dil. NH_3 solution. What is the formula and stereochemical shape of the product? (7)
What use does this reaction find? (8)

Iron

1 Add about 1–2 grams of iron(III) chloride (preferably anhydrous) or iron(III) ammonium sulphate ('ferric alum') to water (25–50 cm^3) in a beaker. Observe the reaction and comment upon the absence of a precipitate. (1)
Use about 2 cm^3 portions of the solution in the following tests.

(i) Test the pH with litmus paper, pH paper, or a pH meter. How do you explain the pH of this solution? (2)

(ii) Add 1 or 2 cm^3 of concentrated (>1 M) sodium carbonate solution. Record carefully your observations and write equations to explain the reactions occurring. (3)

(iii) Add a piece of clean magnesium ribbon or some magnesium turnings. Explain your observations. (4)

(iv) Add dil. NH_3 solution to an excess. Interpret your observations. (5)

(v) Add an excess of solid sodium citrate, tartrate, or EDTA and warm to dissolve the added solid. Now add dil. NH_3 and interpret your observations. (6)

(vi) Add 1 or 2 cm^3 of potassium (or ammonium) thiocyanate solution (0.1–1.0 M). What is the formula of the iron species present? (7)

2 Prepare an aqueous solution of $FeSO_4 \cdot 7H_2O$ (0.5–1.0 M). What species of iron is present in this solution? (8)
Use 2 cm^3 portions of this solution in the following tests.

(i) Test the pH as in 1(i) and compare it with that obtained in 1(i). How do you account for the difference? (9)

(ii) Add dil. NH_3 to an excess and observe the effect of standing or warming the mixture in air. Explain your observations. (10)

(iii) Add potassium thiocyanate solution as in 1(vi) and explain your observations. (11)

Aluminium

If anhydrous aluminium chloride is available begin with experiment 1; if not do experiment 2 which may be performed with aluminium chloride, aluminium sulphate, or potash alum.

1 Cautiously add a little (about 1 g) $AlCl_3$ to about 25 cm^3 of water in a beaker. Account for the violence of the reaction. (1)

2 Prepare a solution of aluminium sulphate hydrate (about 1.0 M) in water. Use 2–3 cm^3 of this solution in the following tests. What species of aluminium is present in the solid aluminium sulphate hydrate? (2) What aluminium species is present in the aqueous solution? (3)

(i) Test the pH of the solution with litmus, pH paper, or a pH meter. Explain the value you observe. (4)

(ii) Add 1 or 2 cm^3 of concentrated ($>$1 M) sodium carbonate solution. Explain your observations. (5)

(iii) Add a piece of clean magnesium ribbon or magnesium turnings; warm the solution if necessary to begin the reaction. Explain your observations. (6)

(iv) Add dil. NaOH slowly to an excess. Write equations to illustrate the reactions occurring. (7)

(v) Add dil. NH_3 solution to an excess. Explain any differences between this reaction and the previous one. (8)

(vi) Add an excess of solid NH_4Cl followed by dil. NH_3 solution. Did you expect NH_4Cl to influence this reaction with NH_3? Explain the contrast between this result and that observed in the experiments on **nickel** [experiment (ii)]. (Perform the nickel experiment now if you have not already done so.) (9)

(vii) Add an excess of solid EDTA (disodium salt of bis[di(carboxymethyl)amino]ethane) and warm the mixture to dissolve the EDTA. Now add dil. NH_3 solution and explain your observations. (10)

Chromium

The tests in 1 below are best carried out using chrome alum (empirical formula $KCr(SO_4)_2 \cdot 12H_2O$) or the hydrated sulphate $Cr_2(SO_4)_3 \cdot 18H_2O$. If chromium(III) chloride is available compare its colour with that of chrome alum. What can you infer from this comparison? (1) What chromium ion is present in the alum? (2)

1 Use 2–3 cm^3 portions of chrome alum or chromium(III) sulphate solution (about 0.5 M) in the following experiments.

(i) Test the pH of the solution and comment upon the result. (3)

(ii) Add 1–2 cm^3 of concentrated (>1 M) sodium carbonate solution and explain your observations. (4)

(iii) Add 1–2 cm^3 of sodium sulphide (or ammonium sulphide) solution (about 1 M). Warm the solution and identify any gas evolved. Write equations for the reactions occurring. (5)

(iv) Add conc. NH_3 solution slowly to an excess. Allow the solution to stand for about 30 minutes. Record and interpret your observations. (6)

(v) Add an excess of solid EDTA (disodium salt of bis[di(carboxymethyl)amino]ethane) and heat to dissolve the solid. Observe the colour of the solution and comment. (7)

Now add dil. NH_3 solution a drop at a time and comment upon your observations. (8)

(vi) Add dil. NaOH solution slowly to an excess followed by 2–3 cm^3 of hydrogen peroxide solution (10 vol) and then warm the mixture. Explain the sequence of reactions. (9)

2 Use 2–3 cm^3 portions of potassium chromate solution (about 0.1 M) in the following experiments. What shape is the chromate ion? (10)

(i) Add 2–3 cm^3 of dil. H_2SO_4 followed by dil. NaOH solution to an excess. Explain the colour changes. (11)

(ii) Add a few drops of dil. H_2SO_4, 3–4 cm^3 of ethoxyethane and 1 cm^3 of hydrogen peroxide. Shake the mixture once and observe. Use your standard textbook to help you interpret this reaction. (12)

(iii) Using a boiling-tube, add an excess (about 10 g) of granulated zinc and enough conc. HCl to ensure continued brisk effervescence for 5–10 minutes. Decant off some of the solution finally obtained into another test-tube and allow it to stand in the air. Compare its colour with that remaining in contact with the zinc. Record the colours of the solutions. What chromium ions are present in these solutions? (13)

Tin

1 Examine the effect of conc. HNO_3 upon metallic tin by warming pieces of tin with nitric acid in a test-tube. What is the formula of the product containing tin and why is tin(IV) nitrate not formed? (1)

2 If tin(IV) chloride (stannic chloride) is available as the anhydrous liquid, perform the tests below in a FUME CUPBOARD. In the author's opinion it is worth buying or making $SnCl_4$ from tin and chlorine as it is a useful chemical to illustrate the reactions of covalent metal halides

($SiCl_4$, $GeCl_4$, $TiCl_4$, VCl_4, etc., react similarly). If only the hydrated tin(IV) chloride is available, however, then the tests described under **aluminium** 2(i) to 2(vi) should be carried out (clues to the answers to the questions on these tests are given by the aluminium answers).

(i) Open a bottle of tin(IV) chloride to the atmosphere. Observe and comment. (2)

(ii) Add a few drops of tin(IV) chloride to some water in a beaker. Notice particularly whether or not there is a precipitate. Explain your observations. (3)

(iii) Prepare a saturated solution of NH_4Cl in a mixture of 4 cm^3 of conc. HCl with 2 cm^3 of water. Add tin(IV) chloride a drop at a time to this solution. Explain the formation of the product. (4)

Bear in mind what ligands are present in the solution and try to relate this reaction to that occurring in (ii).

(iv) Dissolve about 1 cm^3 of tin(IV) chloride in dry toluene (10–20 cm^3). If you have a cloudy solution, what is the cloudiness due to? (5)

Add to 3–5 cm^3 portions of this solution

(*a*) ethoxyethane (ether) a drop at a time,

(*b*) pyridine, or any tertiary amine, a drop at a time or in small amounts,

(*c*) any other ligand of your choice (suggestions: propanone, benzaldehyde, diphenylmethanone, 2,4-dinitrophenylhydrazine).

Why does tin(IV) chloride react with all the compounds in (*a*), (*b*), and (*c*)? (6)

What formulae and structures do you suggest for the products? (7)

Predict whether or not tin(IV) chloride will react with triphenylphosphine (PPh_3) and suggest a likely formula for any product. (8)

Comment upon the relative reactivities of $SnCl_4$ and CCl_4. (9)

Which of these halides do you expect to resemble $GeCl_4$ more closely as regards chemical reactivity? (10)

ANSWERS TO QUESTIONS IN PRACTICAL WORK

Copper

1 $[Cu(H_2O)_6]^{2+}$.

2 $[CuCl_4]^{2-}$.

3 $CuSO_4$ white, we started with blue solution; $CuCl_2$ not formed in the reaction; arrow in equation pointing in wrong direction; H_2SO_4 stronger acid than HCl.

4 $[Cu(H_2O)_6]^{2+}$ produced as H_2O concentration increases. Equilibrium $[Cu(H_2O)_6]^{2+} \underset{H_2O}{\overset{Cl^-}{\rightleftharpoons}} [CuCl_4]^{2-}$.

5 $Cu(OH)_2(s)$ (or basic sulphate); $[Cu(NH_3)_4(H_2O)_2]^{2+}$.

6 Insufficient concentration of NH_3 in conc. NH_3 solution; equilibrium position stops at $[Cu(NH_3)_4(H_2O)_2]^{2+}$.

7 $[Cu(CN)_4]^{2-} \rightarrow [Cu(CN)_4]^{3-}$.

8 $[Cu(CN)_4]^{2-}$, CN^- extreme end of spectrochemical series—transition moved to ultraviolet. $[Cu(CN)_4]^{3-}$ is Cu^I, d^{10} ion; no $d \rightarrow d$ transitions possible.

9 $[Cu(EDTA)]^{2-}$, 1:1 complex formed with very high stability constant; no $[Cu(H_2O)_6]^{2+}$ to undergo exchange with NH_3.

10 White precipitate is $BaSO_4$; therefore precipitate not $Cu(OH)_2$ but basic sulphate.

$$[Cu(H_2O)_6]^{2+} + H_2O \rightleftharpoons [Cu(H_2O)_5(OH)]^+ + H_3O^+$$

$[Cu(H_2O)_5(OH)]_2SO_4$ present in precipitate; amounts and composition depend upon conditions.

11 $[CuCl_4]^{2-} + Cu \rightarrow 2[CuCl_2]^-$; this ion poured into H_2O, concentration of Cl^- thus reduced, and CuCl precipitates. CuCl white as $Cu^I = d^{10}$. In NH_3, $[Cu(NH_3)_2]^+$ formed; again colourless, but oxidized rapidly by oxygen in the air to blue $[Cu(NH_3)_4(H_2O)_2]^{2+}$.

Silver

1 AgCl.

2 $[Ag(NH_3)_2]^+$.

3 $[AgCl_2]^-$.

4 Ag_2O; $[Ag(OH)_2]^-$ in conc. NaOH.

5 $[Ag(NH_3)_2]^+$; Ag_2O may be seen momentarily as a brown precipitate especially if more concentrated silver nitrate solution is used.

6 Ag_2CrO_4.

7 Ag_2CrO_4 dissolves in acid, therefore chromate indicator only useful in neutral solution; $CaCO_3$ usually added to ensure neutrality in silver nitrate titrations.

8 $AgCl$; $[Ag(CN)_2]^-$.

9 $Ag_2S_2O_3 \rightarrow [Ag(S_2O_3)_2]^{3-}$.

10 Photographic fixing.

11 Linear two-coordinate complexes predominantly formed.

Cobalt

1 Blue, $Co(OH)_2$.

2 $Co(OH)_2 \xrightarrow{NH_3} [Co(NH_3)_6]^{2+} \xrightarrow{O_2 \text{ or } H_2O_2} [Co(NH_3)_6]^{3+}$.

3 Blue $Co(OH)_2 \rightarrow$ pink $Co(OH)_2$.

4 Oxidation by air $\rightarrow$ brown '$Co(OH)_3$' probably best described as $CoO(OH)$.

5 $Co(CN)_2 \rightarrow$ green solution of $K_3[Co(CN)_5(H_2O)] \rightarrow$ brown $K_3[Co(CN)_6]$ upon oxidation.

6 $[Co(H_2O)_6]^{2+} + 4Cl^- \rightleftharpoons [CoCl_4]^{2-} + 6H_2O$.

7 $[Co(H_2O)_6]^{2+} + 4NCS^- \rightleftharpoons [Co(NCS)_4]^{2-} + 6H_2O$.
Hg^{2+} precipitates the tetrahedral tetraisothiocyanatocobaltate(II) anion:

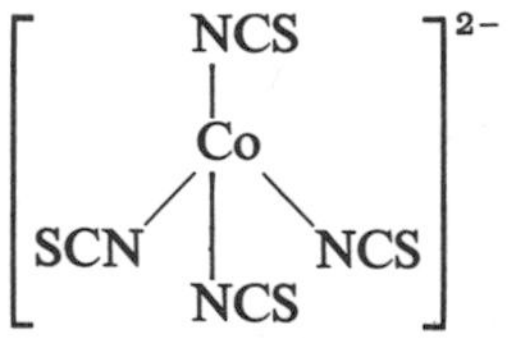

```
⌈       NCS        ⌉ 2-
|        |         |
|        Co        |
|      / | \       |
| SCN    |   NCS   |
⌊       NCS        ⌋
```

8 $[Co(H_2O)_6]^{2+} + 6NO_2^- \rightleftharpoons [Co(NO_2)_6]^{4-} + 6H_2O$; $[Co(NO_2)_6]^{4-}$ oxidized by HNO_2 from $KNO_2 + CH_3CO_2H \rightarrow$ $[Co(NO_2)_6]^{3-}$ which precipitates as $K_3[Co(NO_2)_6]$.

Nickel

1 $Ni(OH)_2$ precipitates; soluble excess $NH_3 \rightarrow [Ni(NH_3)_6]^{2+}$.

2 $[Ni(NH_3)_6]Br_2$.

3 NH_4Cl reduces concentration of OH^-; substitution reaction occurs without hydrolysis of $[Ni(H_2O)_6]^{2+}$; $[Ni(NH_3)_6]Cl_2$ soluble.

4 $[Ni(H_2O)_6]^{2+} + 2H_2O \rightleftharpoons [Ni(OH)_2(H_2O)_4] + 2H_3O^+$
$2H_3O^+ + 2OH^- \rightarrow 4H_2O$.

5 $Ni(CN)_2$; $[Ni(CN)_4]^{2-}$. **6** Square planar.

7

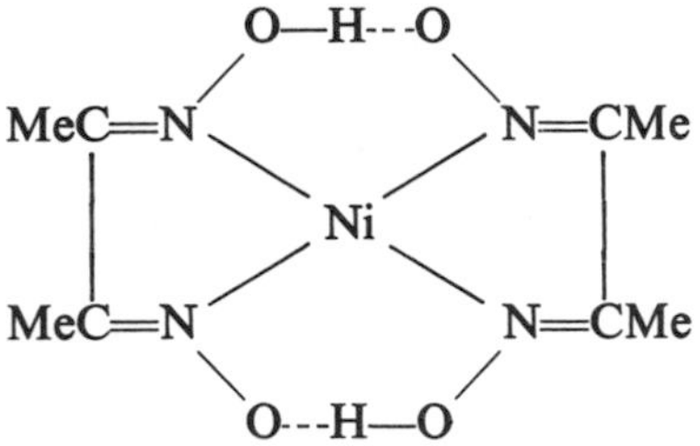

```
          O—H- -O
         /        \
MeC=N               N=CMe
   |     \        /     |
   |        Ni          |     ;
   |     /        \     |
MeC=N               N=CMe
         \        /
          O- -H—O
```

square planar arrangement of 4Ns around Ni.

8 Gravimetric estimation of nickel.

Iron

1 $[Fe(H_2O)_6]^{3+}$ formed (already present in the alum). Hydrolysis to $[Fe(H_2O)_5(OH)]^{2+}$ partially occurs. No base present therefore no further hydrolysis and insufficient olation and oxolation occurring to cause precipitation.

2 $[Fe(H_2O)_6]^{3+} + H_2O \rightleftharpoons [Fe(H_2O)_5(OH)]^{2+} + H_3O^+$; solution acidic since H_3O^+ present.

3 Above equation followed by

$$[Fe(H_2O)_5(OH)]^{2+} + H_2O \rightleftharpoons [Fe(H_2O)_4(OH)_2]^+ + H_3O^+$$
$$[Fe(H_2O)_4(OH)_2]^+ + H_2O \rightleftharpoons [Fe(H_2O)_3(OH)_3] + H_3O^+$$
$$2H_3O^+ + CO_3^{2-} \rightarrow CO_2 + 3H_2O.$$

4 As above $Fe(OH)_3$ precipitated eventually, H_3O^+ removed by Mg:

$$Mg + 2H_3O^+ \rightarrow Mg^{2+} + H_2 + 2H_2O.$$

5 $Fe(OH)_3(s)$ formed. No NH_3 complexes; Fe marked preference for oxygen donors.

6 $[Fe(EDTA)]^-$, $[Fe(cit)_3]^{3-}$, $[Fe(tart)_3]^{3-}$ complexes of chelating ligands very stable; no reaction with NH_3; no $[Fe(H_2O)_6]^{3+}$ present.

7 $[Fe(H_2O)_5(SCN)]^{2+}$. **8** $[Fe(H_2O)_6]^{2+}$.

9 Dipositive ion barely acidic; tripositive ion strongly acidic.

10 $Fe(OH)_2$ oxidizes rapidly in air to brown $Fe(OH)_3$.

11 Fe^{2+}–SCN^- complexes not strongly coloured; pink colour due to $[Fe(SCN)(H_2O)_5]^{2+}$, i.e. delicate test for the presence of Fe^{3+} in original solution.

Aluminium

1 High enthalpy change of solvation of Al^{3+} and $3Cl^-$ ions in reaction:

$AlCl_3 + H_2O(excess) \rightarrow [Al(H_2O)_6]^{3+} + 3Cl^-(aq)$.

2 $[Al(H_2O)_6]^{3+}$. **3** $[Al(H_2O)_6]^{3+}$.

4 $[Al(H_2O)_6]^{3+} + H_2O \rightleftharpoons [Al(H_2O)_5(OH)]^{2+} + H_3O^+$; solution acidic since H_3O^+ present.

5 As above followed by (in two stages)
$[Al(H_2O)_5(OH)]^{2+} + 2H_2O \rightleftharpoons [Al(H_2O)_3(OH)_3] + 2H_3O^+$
$2H_3O^+ + CO_3^{2-} \rightarrow CO_2 + 3H_2O$.

6 $Al(OH)_3$ precipitated as in 5, hydrogen evolved by
$Mg + 2H_3O^+ \rightarrow Mg^{2+} + 2H_2O + H_2$.

7 $[Al(H_2O)_6]^{3+} + 3H_2O \rightleftharpoons [Al(H_2O)_3(OH)_3] + 3H_3O^+$
$[Al(H_2O)_3(OH)_3] + OH^- \rightleftharpoons [Al(H_2O)_2(OH)_4]^- + H_2O$.

8 No aluminate formation; OH^- concentration insufficient (above equilibrium lies to left in low OH^- concentration).

9 NH_4Cl no effect on $Al(OH)_3$ precipitation. Since $[Al(H_2O)_6]^{3+}$ is very acidic only low OH^- concentration required to remove H_3O^+ and force equilibrium to the 'hydroxide' stage. $[Ni(H_2O)_6]^{2+}$ not very acidic so stronger base required to drive hydrolysis to $Ni(OH)_2(s)$.

10 $[Al(EDTA)]^-$ formed, very high stability constant, no $[Al(H_2O)_6]^{3+}$ to undergo reaction with NH_3.

Chromium

1 Chrome alum red violet, $CrCl_3 \cdot 6H_2O$ dark green (normally purchased isomer). Therefore different environment around chromium or different structure. Chrome alum has $[Cr(H_2O)_6]^{3+}$; dark green $CrCl_3 \cdot 6H_2O$ contains $[Cr(H_2O)_4Cl_2]^+$, i.e. $[Cr(H_2O)_4Cl_2]Cl \cdot 2H_2O$.

2 $[Cr(H_2O)_6]^{3+}$.

3 Acidic since $[Cr(H_2O)_6]^{3+} + H_2O \rightleftharpoons [Cr(H_2O)_5(OH)]^{2+} + H_3O^+$.

4 $Cr(OH)_3(H_2O)_3$ precipitates with evolution of CO_2.
$[Cr(H_2O)_6]^{3+} + 3H_2O \rightleftharpoons [Cr(H_2O)_3(OH)_3] + 3H_3O^+$
$2H_3O^+ + CO_3^{2-} \rightarrow CO_2 + 3H_2O$

5 $[Cr(H_2O)_6]^{3+} + 3H_2O \rightleftharpoons [Cr(H_2O)_3(OH)_3] + 3H_3O^+$
$2H_3O^+ + S^{2-} \rightarrow H_2S(g) + 2H_2O$

6 $Cr(OH)_3$ precipitated initially by above reactions. Some solubility to give violet solution of chromium ammines, e.g. $[Cr(NH_3)_6]^{3+}$.

7 $[Cr(EDTA)]^-$ formation observed by colour change.

8 No precipitation with NH_3—no $[Cr(H_2O)_6]^{3+}$ left.

9 $Cr(OH)_3(H_2O)_3$ green–grey, soluble in an excess of OH^-:
$[Cr(OH)_3(H_2O)_3] + 3OH^- \rightleftharpoons [Cr(OH)_6]^{3-} + 3H_2O$;
green chromate(III) oxidized by H_2O_2 to chromate(VI):
$2[Cr(OH)_6]^{3-} + 3H_2O_2 \rightarrow 2CrO_4^{2-} + 8H_2O + 2OH^-$.

10 Tetrahedral.

11 $2CrO_4^{2-} + 2H_3O^+ \rightarrow \underset{\text{orange, dichromate}}{Cr_2O_7^{2-}} + 3H_2O$

$$Cr_2O_7^{2-} + 2OH^- \rightarrow \underset{\text{yellow, chromate}}{2CrO_4^{2-}} + H_2O$$

12 Deep-blue chromium(VI) peroxo compound $CrO(O_2)_2$ formed; more stable in ethereal solution.

13 $Cr_2O_7^{2-} \xrightarrow{HCl} \underset{\text{orange}}{[CrO_3Cl]^-} \rightarrow \underset{\text{green}}{[Cr(H_2O)_6]^{3+}} \xrightarrow{Zn} \underset{\text{blue}}{[Cr(H_2O)_6]^{2+}}$.

(blue → green: readily oxidized by air)

Tin

1 SnO_2; $[Sn(H_2O)_6]^{4+}$ hydrolyses.

2 $SnCl_4$ hydrolyses in moist air.

3 Hydrolysis, $Sn(OH)_4$ may not be precipitated. In high concentration of Cl^- (4HCl produced for every $Sn(OH)_4$) soluble chloro complexes formed, $[Sn(OH)_xCl_{6-x}]^{2-}$, e.g. $[SnCl_6]^{2-}$.

4 $SnCl_4 + 2Cl^- \rightarrow [SnCl_6]^{2-}$; precipitate is $(NH_4)_2SnCl_6$.

5 Hydrolysis; toluene wet!

6 $SnCl_4$ Lewis acid; ethers, etc., Lewis bases.

7 $SnCl_4(Et_2O)_2$; $SnCl_4(py)_2$; $SnCl_4L_2$; octahedral.

8 Ph_3P Lewis base; $SnCl_4(PPh_3)_2$.

9 CCl_4 cannot exceed octet (d orbitals too high in energy); $SnCl_4$ has d orbitals available for bonding so can react readily with ligands, such as H_2O, Cl^-.

10 $SnCl_4$.

Index